Contents

Responsibility for CED Statements on National Policy

The Committee for Economic Development is an independent research and educational organization of two hundred business executives and educators. CED is nonprofit, nonpartisan, and nonpolitical. Its purpose is to propose policies that will help to bring about steady economic growth at high employment and reasonably stable prices, increase productivity and living standards, provide greater and more equal opportunity for every citizen, and improve the quality of life for all. A more complete description of the objectives and organization of CED is to be found in the section beginning on page 120.

All CED policy recommendations must have the approval of the Research and Policy Committee, a group of sixty trustees whose names are listed on these pages. This Committee is directed under the bylaws to "initiate studies into the principles of business policy and of public policy which will foster the full contribution by industry and commerce to the attainment and maintenance" of the objectives stated above. The bylaws emphasize that "all research is to be thoroughly objective in character, and the approach in each instance is to be from the standpoint of the general welfare and not from that of any special political or economic group." The Committee is aided by a Research Advisory Board of leading social scientists and by a small permanent professional staff.

The Research and Policy Committee offers this statement as an aid in bringing about greater understanding of the interrelated policy problems

BROADCASTING and CABLE TELEVISION

POLICIES FOR DIVERSITY AND CHANGE

*A Statement on National Policy
by the Research and Policy Committee
of the Committee
for Economic Development*

April 1975

Library of Congress Cataloging in Publication Data

Committee for Economic Development.
  Broadcasting and cable television—policies for
diversity and change.

    1. Television broadcasting—United States.
2. Community antenna television—United States.
3. Television programs, Public service—United States. I. Title
HE8700.8.C65 1975   384.55'4'0973   75-6536
ISBN 0-87186-758-3 lib. bdg.
ISBN 0-87186-058-9 pbk.

First printing: April 1975
Paperbound: $2.50
Library binding: $4.00
Printed in the United States of America by Georgian Press, Inc.
Design: Harry Carter

Photos: page 11, Bob Peterson, Time-Life; page 27, United Press International;
page 45, Children's Television Workshop; page 59, Del Ankers Photographers;
page 79, Ankers Capitol Photographers; cover, Eli Finer from De Wys.

COMMITTEE FOR ECONOMIC DEVELOPMENT
477 Madison Avenue, New York, N.Y. 10022

facing commercial broadcasting, public broadcasting, and cable television, and of actions necessary for achieving diversity in a period of transition from scarcity to abundance in electronic communications. The Committee is not attempting to pass judgment on any pending specific legislative proposals; its purpose is to urge careful consideration of the objectives set forth in the statement and of the best means of accomplishing those objectives.

Each statement on national policy is preceded by discussions, meetings, and exchanges of memoranda, often stretching over many months. The research is undertaken by a subcommittee, assisted by advisors chosen for their competence in the field under study. The members and advisors of the Subcommittee on the Economic and Social Impact of the New Broadcast Media, which prepared this statement, are listed on page 6.

The full Research and Policy Committee participates in the drafting of findings and recommendations. Likewise, the trustees on the drafting subcommittee vote to approve or disapprove a policy statement, and they share with the Research and Policy Committee the privilege of submitting individual comments for publication, as noted on this and the following page and on the appropriate page of the text of the statement.

Except for the members of the Research and Policy Committee and the responsible subcommittee, the recommendations presented herein are not necessarily endorsed by other trustees or by the advisors, contributors, staff members, or others associated with CED.

6

■ Purpose of this Statement

This statement by the CED Research and Policy Committee is an earnest effort by a group of business executives and educators to come to grips with vital issues of national communications policy. The decision to venture into a field that has long been the domain of specialists stems from an increasing awareness that the electronic media play a critical role in determining the shape of the economy and the society. Citizens rely heavily on television and radio for news, information, and entertainment, and the nation's economic system is greatly dependent on these media for marketing its goods and services.

Our concern about the complexities and delicate aspects of the subject caused us first to establish a task force that met in early 1971 to assess CED's potential contribution to the policy discussion in this field. This group was composed of twenty-eight representatives of business, including the commercial broadcasting and allied electronics fields, public broadcasting, educational and research institutions, and citizens groups. The choice of broadcasting and cable television as a subject of study was given unanimous support by this task force. Moreover, the task force found CED particularly qualified to lend what one member called "a combination of vision and objectivity plus hard business reality."

Although vigorous dissent is documented in footnotes, I believe that this statement, *Broadcasting and Cable Television: Policies for Diversity and Change,* fully validates the decision of that panel. Its proposals focus on commercial broadcasting, public broadcasting, and cable television as three critical elements in an emerging abundance in communications. It is the Committee's view that problems in the communications field are part of a fabric of interwoven policy issues and that the manner in which they are resolved can strongly influence future patterns of American business, education, and entertainment.

An abundance in electronic communications, made possible by cable, satellites, video cassettes, and other advancing technologies, raises large questions about the importance, costs, and benefits to society of a proliferation of media, channels, and voices; the proper degree of government regulation; and the roles that each of the various media should play in serving the public interest. We do not claim to furnish long-term answers to such questions. The strong differences among some members of the Committee illustrate the difficulty of resolving policy issues in this field.

Yet, the Committee has successfully forged a consensus for recommendations that provide at least a sound basis for public discussion.

Recognizing that changes in communications policy will not proceed in an orderly fashion according to a specific timetable, the report suggests certain practical steps that should be taken to ease the difficult transition from the present economic and regulatory policies designed for an era of scarcity of communications channels to policies geared to an era not limited by such constraints.

The recommendations embody several common themes. We emphasize the need for research, analysis, and experimentation during the period of transition. We suggest gradually diminishing government regulation and increasing competition among the technologies as experience about the impact of such changes is accumulated and evaluated. We stress the need for the media to establish goals and specific objectives that will not only help them to determine their place in a national communications system but also provide the public with a yardstick for measuring their performance. Finally, we emphasize the requirement for greater public participation in shaping our communications system.

Specifically, we view over-the-air commercial broadcasting as a means of meeting mass-audience requirements for news and entertainment. We emphasize that it is in the self-interest of commercial broadcasters to be more responsive to changing demands from the public. An enlightened sense of social responsibility should lead the broadcast media to deal voluntarily with controversial issues such as violence on television, thus dampening pressures for government control of program content. Until the advent of an abundance of channels in use lessens the need for regulation, we foresee the need for continued, but monitored, regulation of broadcasters as trustees of the public airwaves.

Public television and radio have the potential of meeting program needs of specialized audiences. But this potential can be realized only if the federal government and private sources provide public broadcasting with the necessary long-term funding on a matching basis to ensure its independence. Accordingly, corporations, foundations, and individuals have a responsibility to contribute funds if they wish to ensure diversity, quality, and independence for public broadcasting. We also suggest improvements in the management of public broadcasting stations and steps by which they can adapt their programming to the new technologies.

Cable television, which can both extend the reach of broadcast signals and generate its own programs, has the capacity for offering the diversity we speak of. Cable should be allowed to compete with over-the-air broadcasting and prove its value in the marketplace. To attract subscribers, we urge policies that will gradually expand the availability of certain programs on cable through both general cable subscription and special pay-cable plans. We also recommend the selective relaxation of regulations prohibiting cable ownership by networks and broadcasters; this would allow firms with the greatest interest in communications to invest the huge amounts of capital needed. Other changes that we propose would alter copyright laws to protect program owners and bring some order out of the present chaos in government regulation of cable systems.

Finally, we urge the federal government to organize and equip itself to accommodate the emerging abundance in communications channels. We recommend relieving the Federal Communications Commission of its judicial burdens and strengthening its research and analysis capabilities.

The CED Subcommittee on the Economic and Social Impact of the New Broadcast Media, which prepared this statement, brought together an extraordinary range of talents, interests, and experience. The task force which met in 1971 formed the core of this subcommittee. Included among the CED trustees were representatives of commercial broadcasting, major television advertisers, representatives of the electronics industry, and a number of university presidents. This group of trustees was supported by others with wide experience in the many specialized subjects that we examined. Among the nontrustee members and advisors were a former chairman and a former commissioner of the Federal Communications Commission, a former director of the United States Information Agency, two former board chairmen of the Corporation for Public Broadcasting, and a number of other experts who contributed greatly to the formulation of this statement. The list of subcommittee members appears on page 6. I acknowledge particularly the persuasive leadership of John L. Burns, a former president of RCA and a former board chairman and chief executive officer of Cities Service Company, who served as chairman of the subcommittee and guided this project to a successful conclusion. He was ably and devotedly assisted by Sol Hurwitz, vice president and director of information for CED, who responsibly discharged the arduous duties of project director.

We are indebted to the John and Mary R. Markle Foundation, the Six Foundation, and the Benton Foundation for their generous contributions to this project.

Philip M. Klutznick, *Chairman*
Research and Policy Committee

An American family watches television. "The expanded use of cable, satellites, video cassettes and other emerging technologies can not only broaden the use of the television set but also free the viewer from the constraints of the broadcast schedule."

1. Introduction and Summary of Recommendations

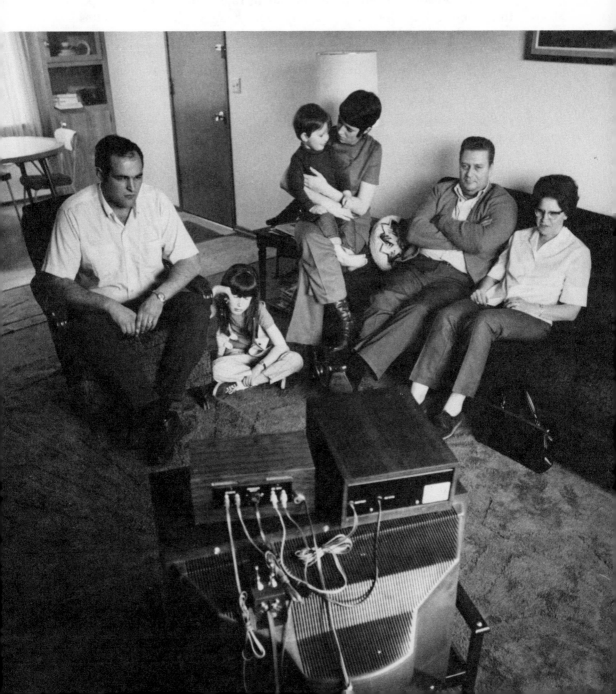

BROADCASTING IS A CENTRAL NERVE of the nation. It transmits information instantaneously and simultaneously to all parts of the country and thus unites the population in a common experience. Broadcasting educates, informs, and entertains. It is also the most effective means yet devised for the dissemination of ideas and the mass merchandising of goods and services. Its influence on public opinion and national values is profound.*

The broadcast media play a dominant role in the way people spend their time and live their lives. Nearly every home in the United States is equipped with at least one television or radio set. In the average home, the television plays more than six hours a day. In fact, television viewing and radio listening together account for the bulk of the average American's leisure time. The images, impressions, and messages conveyed by television and radio help shape the environment and condition the individual's response to the world around him. As business executives and educators, we are deeply impressed with the impact of the broadcast media on our daily experience.

There can be no doubt that the broadcasting industry deserves credit for the enrichment of American life through its development of mass entertainment and information. The enormous popularity of American television programs in other countries is further testimony to its success. Television has become the principal means by which Americans see themselves as a society and a potent force in the nation's political system. It has played a historic role in its coverage of social and political events and continues to render a vital service to the nation.

This creditable performance has sharpened the appetite of viewers for news and information and has aroused increasing demands for greater quality, diversity, and choice in programs. Although public broadcasting has emerged as a medium that can meet many of these growing demands, the lack of adequate long-range funding has stifled its development.

Public policy regarding broadcasting is premised on the fact that broadcasters are trustees of a scarce public resource: the airwaves. As such, they are required by law to operate in the "public interest, convenience, or necessity" and to satisfy the tastes, interests, and needs of the communities they serve. Serious questions have been raised about whether broadcasters are adequately fulfilling these public responsibilities.** But there are equally serious questions about whether the public-interest requirements now imposed on broadcasting (but on no other mass medium) will make sense in a new regulatory environment in which scarcity will cease to be the controlling factor.

See memoranda by *C. WREDE PETERSMEYER and by JOHN A. SCHNEIDER, page 89.
See memorandum by **C. WREDE PETERSMEYER, page 90.

FROM SCARCITY TO ABUNDANCE

Thanks to enormous strides in technology, the nation is entering an era in communications that offers opportunities for access to a new diversity and abundance of electronic channels and voices. Our policy statement anticipates this new era and suggests certain practical steps that should be taken to help ease the transition from a regulatory and economic policy based on scarcity to one that is responsive to abundance.

The airwaves are a scarce public resource because of the technical limitations imposed by the use of the electromagnetic spectrum. But scarcity can be defined in many ways, not merely in terms of the limited number of channels that are available. After all, there are far more radio and television stations in this country than there are daily newspapers. Scarcity can also be defined in terms of supply and demand for spectrum space. Broadcasting occupies the very choice portions of the spectrum at a time when technology is creating increasing demands for spectrum space from nonbroadcast users in industry, finance, and government. In fact, new users are now crowding the spectrum faster than technology is finding ways to accommodate them. Until recently, the opposite had been true.*

But technology has also found ways to bypass this bottleneck. The expanded use of cable,[1] satellites, video cassettes, and other emerging technologies can not only broaden the use of the television set but also free the viewer from the constraints of the broadcast schedule. Thus, every home, school, and business can gain access to a nearly unlimited number of channels for a multiplicity of uses: commercial, public, and community television and radio; two-way communications for banking and shopping; facsimile reproduction; and a host of services involving the storage, transmission, and retrieval of information. Satellites can widen opportunities for economic program distribution by joining together diverse media to meet common functional and geographic needs. In the

---

[1] Because the term *cable* is deceptively imprecise, policy issues are often obscured by confusion over definitions. The word is often used interchangeably with other terms to denote the multichannel capacity of coaxial cable: *cable TV, CATV (community antenna television), broadband distribution systems,* and *coaxial communications.* Coaxial cable is only one form of broadband communications technology. Although it is the one dealt with almost exclusively in this statement, the recommendations dealing with cable could also apply to other broadband communications technologies. See "New Communications Technologies: A Brief Guide," page 23.

See memorandum by ·C. WREDE PETERSMEYER, page 91.

future, they can provide the universal link for broadcasting and related technologies. In short, the television set can become a receiver of information in endless quantities and varieties, programmed at the user's convenience.

The technology is at hand to effect a radical readjustment of present patterns of industry, finance, education, entertainment, and leisure. Ultimately, any person will be able to have instant audiovisual communication with any other person throughout the world. The question is no longer whether such a result is technically possible but whether it is economically feasible and socially desirable. In 1854, Henry David Thoreau wrote: "We are in great haste to construct a magnetic telegraph from Maine to Texas; but Maine and Texas, it may be, have nothing important to communicate."[2] In an era of instant communication, in which everyone can talk to everyone else, in which every thought, experience, and event can be transmitted, stored, and retrieved, the biggest challenge may be deciding what to communicate.

AN INFORMATION SOCIETY

This Committee recognizes that knowledge and information are major factors in the growth and productivity of the American economy and in the well-being of the society. Advances in communications technology can make information accessible more quickly and in larger quantities and can provide means through which more of the public can take part in the creative process of communications.

Will a proliferation of electronic media also assure greater selectivity and quality of information? People are limited in the amount of information they can absorb. If the move from scarcity to abundance in communications does not guarantee better or more complete information, if it only guarantees *more*, then it may well serve no constructive purpose. It is doubtful that our society needs a multitude of media beaming identical or similar messages to a scattered and fragmented audience. Moreover, in an era of explosive growth in communications, enlarging the reach and impact of the media may encroach on individual freedom and privacy.*

---

[2] Henry D. Thoreau, *Walden*, ed. J. Lyndon Shanley (Princeton: Princeton University Press, 1971), p. 52.

See memorandum by *C. WREDE PETERSMEYER, page 92.

Abundance of communications, therefore, is relative; it is measured by its cost and its benefits to society. Although it is essential to continue to explore the frontiers of communications, their practical limits must be known and understood.

The future of electronic communications will be shaped, not by technology alone, but by society's needs tempered by harsh economic realities. The critical questions are:

> In the allocation of the nation's resources, how important is the goal of virtually unlimited channels of communication?
>
> How will the capital and creative demands of so many channels be met?
>
> What arrangements best assure that their costs will be matched by their benefits to society?
>
> With what degree of government regulation, through what media, and at what price should citizens get information?

In confronting these and other questions, strong differences emerged within the Committee. We do not pretend to have answered these questions fully in this statement, nor do we believe they are intrinsically of such a nature that answers are readily available elsewhere. Nevertheless, we hope that by asking the right questions and by encouraging research, analysis, and experimentation, we are helping to build a coherent framework within which discussion can take place, answers can be found, and decisions can be made.

There have been few attempts, in government or elsewhere, to determine how broadcasting and the various other communications technologies will fit together. How will they help to shape the kind of society that the United States is likely to be a decade or two hence, and how will they conform to that society? The Communications Act of 1934, the statute that governs broadcasting, did not even anticipate television as it exists today, much less the bewildering array of related technologies that have since emerged. Broadcast policy has been slow to adapt to change, hesitant to cope with new issues. Rules and regulations have often evolved in a random, haphazard manner, without consideration of their long-term implications.

Because of the interdependence of the various parts of broadcasting and the emerging communications technologies, it is difficult to consider any one part of the problem without examining the total system. In order to keep our task to manageable proportions, we have focused most of this statement on policies affecting commercial broadcasting, public broadcasting, and cable television. Our hope is that by concentrating on these three key areas we can help establish the basis for a more comprehensive national program that will bring the many other elements of telecommunications into a unified system.

We do not believe that there is a once-and-for-all solution to the complexities of public policy for broadcasting and the changing communications technologies; certainly, no such solution is available at this stage. Therefore, our recommendations constitute an interim policy, a bridge to the future.

SOCIAL AND ECONOMIC FORCES

Effective public policies for broadcasting and the new technologies cannot ignore the social and economic forces that are helping to shape the society.

As public problems become more complex, citizens are seeking more information and a larger voice in the critical decisions that affect them. Broadcasting and the new technologies can widen the opportunities for more informed and more direct citizen involvement in the decision-making process.

Knowledge is becoming increasingly perishable, bringing new demands for retraining, career advancement, and adult education. New communications technologies can greatly enhance educational opportunity, productivity, and effectiveness through a variety of open learning systems not bound by buildings or campuses.

Social and demographic patterns are shifting. Increasing numbers of married women with children are entering the labor force. At the same time, the elderly constitute a growing portion of the population. Shorter workweeks and earlier retirements are expanding leisure time. Such changes in the makeup of American society have significant implications for television programming and viewing.

In the formulation of public policy, these and other trends pose policy issues and require study and analysis that go well beyond the scope of this statement. We urge a continuing and systematic examination of the consequences of major social and economic developments as a prerequisite for effective policy making.

FIVE POLICY IMPERATIVES

In developing recommendations for public policy, the Committee was guided by these five imperatives:

1. In the transition from scarcity to greater abundance and diversity, broadcast policy should rely more on competitive market forces and less on government regulation. Fair competition among the technologies should be encouraged.* The development of new communications media should be fostered by freer access to the marketplace.

2. Public broadcasting can make an important contribution to greater choice and quality of programs. To assure the strength and independence of public broadcasting, adequate long-range financing is essential.

3. Cable's broadband capacity has the potential for extending television and offering access to an abundance of communications services. Cable should be allowed to prove its value in the marketplace. * †

4. Opportunities for new sources of talent and creative programming should be encouraged. Only by enhancing quality, diversity, and accessibility of programs can a greater abundance of media serve the public good.

5. The organization and management of the federal government's communications responsibilities must be modernized to permit development of more coherent and responsive national policies.

From our recommendations, the general outlines of a national telecommunications system begin to emerge. In this system, commercial over-the-air broadcasting meets the mass-audience requirements for news and entertainment, and public broadcasting fills the varied needs of specialized audiences by providing greater choice, diversity, and enrichment. Cable complements over-the-air commercial and public broadcasting by improving its signal and extending its reach and by offering, through an abundance of channels, greater opportunities for specialized programs and nonbroadcast services. Free, subscription, and direct payment mechanisms should coexist and compete, just as they do in the print media.

Long-range government policy must be fashioned accordingly. It must promote the public interest in diversity and fair competition in an environment no longer controlled by spectrum scarcity.

See memorandum by *C. WREDE PETERSMEYER, page 92.
See memorandum by †ROBERT R. NATHAN, page 92.

We are framing our proposals against a rapidly changing panorama of political, social, economic, and technological developments. Therefore, a serious need for research and experimentation underlies all our policy imperatives.

Centers of high-quality research should be established and nourished both within and outside the government. These institutions should be devoted to improving understanding of the impact of telecommunications on society and to projecting the implications of alternative public policies. Industry, education, government, and of course, the media themselves have a vital stake in the future of electronic communications. They bear a large responsibility for applying their talent, resources, and creative energies to finding solutions to national policy questions.

Finally, no improvements can take place without strong popular support. The viewer and listener are commonly considered the ultimate arbiters of broadcast policy; they alone hold the power to turn the set on or off and to switch from one channel to another. But it is not enough merely to exercise choice from among the available options. More effective public participation is needed to help determine what those options ought to be.

Summary of Recommendations

The recommendations summarized here are discussed in greater detail in subsequent chapters.*

Commercial Broadcasting and the Public. The public's expectations of what commercial broadcasting should contribute to society have risen markedly over the years. We believe enlightened self-interest requires that commercial broadcasting upgrade its social performance in line with this new public awareness. As trustees of a scarce public resource, whose activities are broadly regulated by the federal government, commercial broadcasters should be doubly conscious of their social role. A number of public-interest requirements have evolved from the concept of the broadcaster as public trustee. For example, broadcasting is covered by a *fairness doctrine* that requires the presentation of contrasting viewpoints on controversial issues of public importance and by a law that insists on *equal time* for all competing political candidates. A new era of abundance and diversity should lessen the need for such federal controls. But mean-

See memorandum by *OSCAR A. LUNDIN, page 93.

while, there are measures that should be taken to help broadcasters strengthen their social performance.

Self-monitoring is also vital. The ways in which commercial broadcasters deal with issues such as the portrayal of violence on television may establish a pattern for resolving conflict and adjusting to changing social needs *voluntarily,* without censorship or intrusion from government or special-interest groups.

FAIRNESS DOCTRINE. As a temporary safeguard, we believe that the fairness doctrine should be maintained in its present form.* The Federal Communications Commission should continue to rule on fairness complaints promptly so that they can be disposed of while the record is fresh and so that any deficiencies can be corrected.* However, the fairness doctrine should be reviewed periodically. When an abundance of electronic channels permits a large-enough number and variety of voices to assure the airing of many viewpoints on controversial issues, the fairness doctrine should be discontinued. In moving toward this goal, we recommend that the Federal Communications Commission authorize limited experiments in which the fairness doctrine would be suspended.*

EQUAL TIME. As a first step toward total repeal of Section 315 of the Communications Act, the so-called equal-time provision, we support elimination of the provision only for candidates for president and vice-president. Meanwhile, Congress and the Federal Communications Commission should develop and periodically review criteria by which broadcasters might allocate free time to candidates for congressional, state, and local office with a view toward complete repeal of the equal-time requirement when the increase in the number of channels indicates that it is no longer warranted.*

CONGRESSIONAL BROADCASTS. Broadcasters should be allowed to transmit by television and radio important events taking place on the floors of the Senate and the House and in committee hearings, subject to rules established by Congress in consultation with representatives of the broadcasting industry.

Public Broadcasting. We believe that public broadcasting can provide greater quality, diversity, and choice in programming. Thus, it can serve the many specialized audiences that commercial broadcasting, with its mass appeal, is not geared to reach. Public television and public radio are in critical need of reliable long-range funding, but we consider recent government proposals for such funding inadequate. At the same time, we recognize that the success of any proposal for increased long-term funding

See memoranda by *C. WREDE PETERSMEYER, pages 93 and 94.

will depend on the establishment of realistic and firm goals for the future of the system. Public television and radio must identify and address the needs of their audiences. Increased resources must be effectively managed. Moreover, public broadcasting must examine how it will fit into an era in which cable and other technologies could substantially widen program diversity and choice.

LONG-RANGE FINANCING.   Support of public broadcasting from general federal tax revenues should be authorized and appropriated by Congress for a period of no less than five years. The level of federal support for public broadcasting in any fiscal year should match nonfederal support on a one-to-two basis up to an established ceiling based on realistic costs of providing expanding quality broadcasting service.

IMPROVING MANAGEMENT.   The Corporation for Public Broadcasting, in collaboration with appropriate business and professional organizations, should provide local television and radio stations with a comprehensive program for improved management, including opportunities for management training and standardized models of budgeting and accounting procedures as well as guidelines for their local application. The implementation of improved management methods should be a major responsibility of local station managers and boards of trustees.

ADAPTING TO NEW TECHNOLOGIES.   To assure a place for public broadcasting among the new technologies, public broadcasters should focus their efforts on programming for a wide variety of purposes: public television stations, commercial stations, cable systems, schools, and individual users. To accomplish this, we urge them to consider expanding their stations into local and regional public telecommunications centers. We also urge public broadcasters to plan now to adapt to other new technologies such as satellites and to offer special services for the deaf and the blind.

Cable Television. Coaxial cable's broadband technology offers the potential of widening the selection of channels far beyond anything presently available through broadcast television.* The basic product of cable is over-the-air commercial and public broadcasting programs whose signals are picked up from local and distant stations and transmitted with improved reception by cable systems to their subscribers. But cable is increasingly originating its own programs or leasing channels to others for that purpose. The cable industry envisions not only a diverse broadcast service with improved signal quality but also a wide selection of non-broadcast entertainment and information services. But this is not now

See memorandum by *C. WREDE PETERSMEYER, page 95.

economically feasible. Consequently, cable is offering mass-appeal programs that compete directly with over-the-air television. The key policy question is: To what extent should cable be allowed to compete with broadcast television for programs and audiences?

We believe that if cable is to become a significant means of widening the range of programming and information services available to the American consumer, it should be allowed to prove its value in the marketplace. The cable industry's short-term need is to obtain as much program material as possible to fill cable's multichannel capacity. This is the only way that cable owners can attract sufficient numbers of subscribers in urban markets to generate the profits necessary to obtain large-scale risk capital.*

Thus, a policy that allows the cable owner to program a minimum number of channels on a nondiscriminatory basis is imperative. For the same reasons, we support the selective relaxation of restrictions on ownership and a gradual phasing out of certain restrictions on pay-cable operations. A strengthened copyright law that will protect program owners and creative artists whose works are made available to cable subscribers is badly needed and long overdue. Measures should also be taken to clarify the conflicting and overlapping federal, state, and local jurisdictions that cloud the regulatory environment in which cable operates. Our recommendations for cable are closely interrelated and must be viewed as an integrated program.

CABLE OWNERSHIP AND ACCESS. **Cable systems should be governed by a policy of nondiscriminatory access. Owners should be allowed to originate or control programming on a limited number of channels but should be required to demonstrate affirmatively that they are not restricting the competitive access of others.**

Restrictions on ownership of cable systems by broadcasters and networks should be relaxed to allow common ownership in selected markets in which a diversity of media and media owners already exists. **

COPYRIGHT. **Congress should modernize and strengthen the copyright law, making it applicable to the retransmission of programs picked up by cable systems from distant broadcast signals.** ***

PAY CABLE. **Programming restrictions on motion pictures and series programs should be phased out gradually and selectively. The Federal Communications Commission should authorize and carefully monitor experiments designed to evaluate the impact of such deregulation on free over-the-air television service. If loosening program controls on movies and series programs leads to unfair competition or other developments**

See memoranda by *ROBERT R. NATHAN , page 92, and by C. WREDE PETERSMEYER, page 95.
See memoranda by **ROBERT R. NATHAN and by C. WREDE PETERSMEYER, page 96.
See memoranda by ***E. B. FITZGERALD and by C. WREDE PETERSMEYER, page 96.

injurious to the public interest, the Federal Communications Commission should take steps to curb such practices; these steps might include re-imposition of the present controls.* Antisiphoning restrictions on major sports events should be maintained, but pay-cable regulations should be modified to allow the presentation of games that are not regularly tele-vised.**

TWO-TIER REGULATION.   We recommend a two-tier system of gov-ernment regulation of cable involving the federal government and the states. The Federal Communications Commission should establish the jurisdictional framework to assure that state regulation is consistent with national policies. State governments should establish state commissions with authority over cable-franchising activities and procedures. Where local conditions warrant, states should delegate franchising powers to local governments, particularly large cities with proven resources for the regulation of a cable system.

Improving the Federal Communications Commission. The Fed-eral Communications Commission (FCC) has had great difficulty formu-lating coherent and responsive communications policy.*** This weakness has two major causes: the increasing scope, complexity, and detail of regu-lation and the lack of time, staff, and budget to develop and execute sound policy. As a result, the commission has had to rely instinctively on tradi-tional rules and precedents that tend to incorporate past practices into policy without reflecting social or technological change. Our recommenda-tions place particular emphasis on the need to relieve the FCC of its judi-cial burdens and on the importance of improving its interdisciplinary re-search and analysis capabilities.

COMMUNICATIONS COURT.   As a means of relieving the Federal Communications Commission of its heavy judicial burdens and allowing it more time for policy formulation, we recommend that the adjudicatory functions now exercised by the commission be conferred upon a new com-munications court.

RESEARCH AND ANALYSIS.   To strengthen its research and analysis capabilities, we recommend that Congress and the Federal Communica-tions Commission give high priority to the growth and development of the Office of Plans and Policy. Funding should be sufficient to allow ex-perienced economists, engineers, attorneys, and social and political scien-tists to provide strong policy research and analysis both for the commis-sion as a whole and for the individual bureaus.

See memoranda by *CHARLES P. BOWEN, JR., and by EDWARD N. NEY, pages 97 and 98.
See memoranda by **CHARLES P. BOWEN, JR., and by E. B. FITZGERALD, page 99, and by C. WREDE PETERSMEYER, page 100.
See memorandum by ***C. WREDE PETERSMEYER, page 100.

NEW COMMUNICATIONS TECHNOLOGIES: A Brief Guide

The rapid developments taking place in communications have brought with them a new lexicon of technical and semitechnical terms. In the jargon that has resulted, technical terms have sometimes been applied in an imprecise manner. A review of some of the phrases currently in use may assist the reader in sorting through the various techniques currently under discussion.

Broadband Distribution Systems permit the distribution of many television, sound, or data channels through a single transmission system. Because each separate channel occupies a defined amount of bandwidth, the simultaneous transmission of many channels requires the bandwidth to accommodate them. *Voice* channels require more bandwidth than telegraph or teletype but have been transmitted in large groups by wire, cable, and microwave techniques for many years. *Television* signals require the greatest bandwidth and make the largest demands on transmission systems. *Data* channels come in bandwidths ranging from smaller than those of voice to equal to or larger than those of television. Broadband systems, then, are usually designed so that large numbers of simultaneous television channels are transmitted with many data, voice, and other channels accompanying them.

Coaxial Cable systems are the most common form of broadband transmission systems. A coaxial cable is a single wire surrounded by a cylinder of metal. The single wire is located at the center of the cylinder and, hence, is coaxial with the cylinder. Dozens of television channels, hundreds of voice conversations, or thousands of low-speed data channels can be transmitted by coaxial cable. Because the systems are relatively inexpensive, coaxial cable can be used for wide-area, high- to medium-density distribution, such as *cable television* or industrial and educational information systems.

Video Recording and **Video Playback** systems have been in use for more than twenty years. Such systems record and play back conventional television audio and visual signals. The current work in this technology is aimed at making them easier and less expensive to operate. There are different types of video recording systems, but all have the same objective of allowing a simple attachment to a home, office, or classroom television set either to play or to record and play back audiovisual programming.

Magnetic Recorders, either **Reel to Reel** or **Cassette,** use the oldest successful video recording medium, *magnetic tape.* Such systems are roughly comparable to audio tape recorders, which use either tape on reels or tape contained in plastic cassettes. The advantages of magnetic video recording include good picture and sound quality, the ability to rerecord on the tape, and the possibility of editing taped material. One disadvantage is the difficulty of making many copies of a single recording; duplication requires special equipment to be economically feasible.

Video Disc systems record television material on *plastic discs* similar to phonograph records. Potentially, such systems can be a low-cost way of producing television programs. However, most video disc systems must mass-produce their product to be economically sound, and the users of such systems cannot record their own video discs.

Microwave systems are broadband communications systems that use ultrahigh-frequency (UHF) *radio signals* to transmit data, telephone communications, and television signals. Although microwave links make up the backbone of the national telecommunications system, it is likely that the longest, most heavily used routes will gradually be replaced by *satellite* channels.

Lasers are specialized devices that generate spectrally pure, high-intensity light beams. Such pure (or coherent) light beams can be modulated with information over a very great bandwidth. However, they are adversely affected by bad weather when transmitted through the atmosphere.

Fiber Optics, when combined with *lasers,* make a very broadband communications channel. A fiber optic channel is a thread or bundle of threads of very pure and carefully fabricated glasses. Such a thread can conduct light very efficiently over long distances. Lasers serve as the source of pure modulated light signals that the fibers can transmit. Tens of thousands of channels can be accommodated by optical fiber systems. This technology is still in the development stages, and it may be 1980 before commercial applications are common.

Communications Satellites orbit the earth as repeaters and amplifiers of signals beamed to them from the ground. They have the advantage of permitting very long distance transmission with very little degradation of the original signal quality. In addition, many receiving antennas on the ground can be aimed at the satellite, allowing widely separated users to receive the signal at the same time and at a relatively low cost. Compared with coaxial cables, satellites offer the potential of reduced cost per channel mile when the transmission path is long. Satellites offer much less bandwidth than the new fiber optic technology, but they allow low-cost transmission for very great distance over either land or water.

Digital Television converts the conventional television signal into a digital data stream formed by a process that encodes the television signal in a prescribed manner. The advantage to *digital encoding* is its resistance to distortion in a long transmission path. Conventional television transmission systems degrade the picture at each point of amplification; digital systems are largely immune to such distortions. Digital encoding can be used with several transmission systems, but because of the complexity of the electronics required, it is unlikely that it will be used for broadcast into homes in the near future.

Computers are excellent tools for controlling complex communications systems. The new electronic telephone exchanges, for example, are based on the use of digital computers to make switching decisions in response to subscriber orders dialed into the computer from telephones. The time may come when it is possible for each householder to use centrally located computers for such matters as banking and ordering of goods and services. Business uses are much more complex because of the much higher volumes of data that are processed.

Senator John F. Kennedy views an offstage monitor as Vice-President Richard M. Nixon appears on television during their third debate in the 1960 presidential campaign. "Broadcasting is the only communications medium covered by a fairness doctrine that requires balance in the presentation of controversial issues of public importance and a law that insists on equal time for competing political candidates."

2. The Public Responsibilities of Commercial Broadcasting

Growing public expectations of broadcasting require a searching examination of the broadcaster's role in society and a more careful determination of how broadcasting can better serve the needs and interests of its audiences. In this chapter, we recommend measures that can help broadcasters to fulfill their public responsibilities more effectively. These measures involve changes in present government regulations as well as means to promote voluntary self-regulation. We must emphasize, however, that the government regulation discussed here is only an interim stage in the process by which broadcasting, like the print and film media, will become virtually free of government controls. As the electronic media move from an era of technological scarcity to an era of abundance—that is, as new ways of reaching the public and new opportunities for ownership of communications systems emerge—the need for government regulation should diminish accordingly, and steps should be taken to ensure that government regulation is, in fact, reduced.*

GOVERNMENT REGULATION OF BROADCASTING

In a democracy, the media perform the special function of disseminating ideas and information that help citizens to exercise free choice. Freedom of the press is one of the philosophic cornerstones on which this nation was founded, and it is carefully safeguarded by the Constitution. The First Amendment provides that "Congress shall make no law . . . abridging the freedom of speech, or of the press. . . ." Certain laws forbidding obscenity, pornography, libel, slander, and criminal incitement are considered to be consistent with the First Amendment when held within bounds set by the courts.

What distinguishes broadcasting from other media has been the scarcity of usable frequencies or channels. By assigning frequencies for private use and enabling the licensee to control both the transmission facilities and the programming, the government has granted broadcasters considerable power over the flow of information. To limit such power, the Communications Act of 1934 provided the FCC with broad authority that inevitably led it to delve into issues involving types and proportionate amounts (and in some cases the content) of programming. Consequently, broadcasting is the only communications medium covered by a fairness doctrine requiring balance in the presentation of controversial issues of public importance and a law that insists on equal time for com-

See memorandum by *JOHN A. SCHNEIDER, page 101.

peting political candidates, however many and however minor. No other medium is required or even encouraged to offer certain categories of public-interest programs as a condition of having licenses awarded or renewed. Indeed, the other media are not licensed at all.

The maze of statutes, rules, regulations, and procedures that govern the conduct of the broadcast media represents an honest attempt by government to define, promote, and protect the public interest. But on occasion, the vast licensing and regulatory powers of government have collided head on with broadcasters' First Amendment rights. It is not entirely clear that the outcome has always served the public interest.

FAIRNESS DOCTRINE

Fairness is one of the most elusive goals of broadcast regulation. The fairness doctrine stems directly from the early notion of the broadcaster as a public trustee (see "The Broadcaster as Public Trustee," pages 30 and 31). The doctrine, which was enunciated as a formal policy by the FCC in 1949, holds that a broadcaster must devote a reasonable amount of time to controversial issues of public importance and that, when he presents one side of such a controversy, he must afford a reasonable opportunity for the presentation of contrasting views.

The fairness doctrine, unlike the equal-time provision for political candidates, allows the broadcaster considerable latitude in making judgments. He is required to offer only *reasonable opportunity* for the discussion of contrasting viewpoints, not a mathematically balanced presentation of every issue. In fact, it has been persuasively argued that the goal of fairness is more likely to be achieved by vigorous advocacy and the presentation of a diversity of viewpoints than by careful attempts to be fair, which frequently succeed in being only bland. The various sides of an argument often do not have equal merit.

Before 1962, the FCC normally reviewed complaints of lack of fairness only at the time of license renewal, when it could examine the broadcaster's overall performance. But in 1962, the FCC decided to deal with fairness complaints as they arise. Now, some thoughtful observers are proposing a return to review only at license renewal.[1]

---

[1] See Henry Geller, *The Fairness Doctrine in Broadcasting: Problems and Suggested Courses of Action* (Santa Monica, Calif.: Rand Corporation, 1973).

The commission has been confronted with fairness questions covering a wide spectrum of material ranging from news and public-affairs programs to commercials and political broadcasts. Its rulings have raised serious questions concerning the criteria for fairness. What is reasonable

THE BROADCASTER AS PUBLIC TRUSTEE

The standard of the public interest was formally introduced into broadcasting by the Radio Act of 1927. The act proclaimed that the airwaves belong to the people and are to be used by individuals only with the authority of short-term licenses granted by the government in the "public interest, convenience, or necessity." The federal government would allocate broadcast channels but could not give away their ownership. Licenses were to be issued for limited periods, "and no such license shall be construed to create any right, beyond the terms, conditions, and periods of the license."

In drafting the 1927 act, Congress could have specified that broadcast frequencies be auctioned to the highest bidder or allocated to users either for a rental fee or with the requirement that a specified amount of a station's time and facilities be devoted to a particular public purpose. Instead, it chose a system of short-term licensing and imposed upon the broadcaster the obligation to operate in the public interest. When one or more applicants competed for a given frequency assignment, the public-interest concept was to become the guide for determining the final award.

The act established the broadcaster as a public trustee, and it is from this concept that public-interest requirements evolved. Regulation was geared to assuring not only equality of reception but equality of transmission service as well. Broadcasters were obligated to devote a reasonable amount of time to airing controversial issues of public importance and to presenting those issues with fairness. Sponsors of broadcast programs were required by law to be identified. Government censorship was forbidden, but the broadcast of obscene or profane material was made a crime.

As a practical matter, evaluating a broadcaster's public performance has not been easy. The commercial broadcaster is licensed balance in the presentation of opposing views? How does a government agency determine whether a viewpoint is favorable, unfavorable, or neutral? In what amount of time, at what hour, and to what audience should an opposing viewpoint be presented? The larger question is: Should the

to use the airwaves for a profit, and in order to maximize his profit, he seeks to attract and hold a mass audience. Understandably, entertainment and sports programs dominate the broadcast schedule and command most of the attention of viewers and listeners.

At the same time, FCC policy on programming makes clear that the "principal ingredient of [the licensee's obligation to operate his station in the public interest] is a diligent, positive, and continuing effort by the licensee to discover and fulfill the tastes, needs, and desires of his [community or] service area," for broadcast service.[2] That policy identifies a wide variety of program categories a station should carry as the major elements usually necessary to meet the public interest, needs, and desires of the community in which the station is located. These include news and public-affairs programs, educational and religious programs, programs for children, weather and market reports, political programs, and editorials (except in the case of noncommercial stations). The FCC does not require a broadcaster to offer a particular percentage of programming in any of these categories, but it does state that a television station must offer a balanced schedule that meets the needs and interests of all substantial groups in its audience. Radio stations, because of their greater number in large cities, are allowed a measure of specialization in their programming.

As a means of developing a more solid basis for renewing licenses, the FCC has periodically attempted to measure a broadcaster's public-service performance in terms of proportions of certain types of programs offered and even in terms of the number of peak viewing hours devoted to such programs. Thus far, these efforts have produced few concrete results.

[2] *Federal Register,* 3 August 1960, p. 7294.

government be involved in such matters at all? And if so, how extensively?

The number of fairness complaints has risen dramatically in recent years, and each requires a response from the broadcaster. But because of limitations of staff, the FCC is often slow in processing them. Moreover, it has failed to develop broad guidelines that would help its staff and the industry dispose of fairness questions more expeditiously.

We recognize the risk that the fairness doctrine may be an inhibiting influence on broadcast journalism. The commission should be extremely sensitive to this risk and should therefore guard against substituting its judgment for journalistic decisions made in good faith. On balance, however, we believe that the fairness doctrine enhances rather than inhibits freedom of speech and of the press by promoting broad discussion of the public's business and vigorous debate on controversial issues of public importance. **As a temporary safeguard, we believe that the fairness doctrine should be maintained in its present form.* The Federal Communications Commission should continue to rule on fairness complaints promptly so that they can be disposed of while the record is fresh and so that any deficiencies can be corrected.***

We expect the situation to be different in the future, however. A central conclusion of this statement is that future abundance and diversity in electronic communications will minimize the need for government regulation and permit a wider play of free-market forces. We believe that the safest guarantee of fair and balanced programming is a policy that allows many voices to compete freely in the marketplace of ideas and to serve a diversity of individual interests, tastes, needs, and opinions. Carefully controlled experiments would help to determine when such a policy should take effect.

The fairness doctrine should be reviewed periodically. When an abundance of electronic channels permits a large-enough number and variety of voices to assure the airing of many viewpoints on controversial issues, the fairness doctrine should be discontinued. In moving toward this goal, we recommend that the Federal Communications Commission authorize limited experiments in which the fairness doctrine would be suspended.* For radio, experiments might be authorized in test markets selected from those that have measurable audiences for fifteen or more AM stations; for television, in test markets selected from those that have measurable audiences for five or more very-high-frequency (VHF) stations.

See memoranda by *C. WREDE PETERSMEYER, pages 93 and 94.

EQUAL TIME

As a political instrument, no mass medium is more powerful than broadcasting. Television and radio exposure have helped elect many candidates to office and have thus established a vital link between the government and the people. Still, certain policies prevent broadcasters from increasing their contribution to the democratic process.

The use of television in political campaigns was examined by this Committee in its 1968 policy statement *Financing a Better Election System*.

> There can be no doubt that the emergence of television has had profound significance in the conduct of political campaigns at all levels. Whether for good or ill, this medium has become the main—and often the decisive —means of communication between candidates and their constituencies for all national, most state, and many local offices.... Any possibility of monopolistic manipulation or inequitable access to this medium would, therefore, constitute the gravest kind of danger to our democratic political system. Fairness in its use is an obvious imperative, and the public interest must be the dominant concern. [page 41]

Fairness remains an obvious imperative, but instead of relying on the general approach of the fairness doctrine, broadcasters are required by Section 315 of the Communications Act of 1934, the so-called equal-time provision, to grant equal time to all competing candidates for a particular public office. The law provides that each station granting time to one candidate must grant equal time to each of his opponents. Exempt from this law are news broadcasts and regularly scheduled news interviews, documentaries (if the appearance of the candidate is incidental to the program), and on-the-spot coverage of events. The result has been that candidates have come to depend almost entirely on paid advertising (a major factor in total campaign costs) in the absence of free debates or other balanced public-affairs programs. Broadcasters have said they would be willing to offer free public-service time to two or three candidates, but the growing number of primary candidacies and minor parties has often made the requirement of equal time for all candidates impossible to meet.

In *Financing a Better Election System*, we noted that in practice equal time is likely to mean "no time at all, since that will avoid disputes over time slots and the troublesome problems of multiple or minor party candidacies. Consequently, Section 315 takes away an opportunity for the major party candidates to communicate with the electorate, without increasing opportunities for minority candidates" (pages 41–42). We maintain, as we did in 1968, that repeal of Section 315 would serve the public interest.

We nevertheless recognize important differences in the way equal time affects candidates at the various levels of government. Whereas the relatively small number of candidates for president and vice-president— and their high popularity and visibility—would make it difficult for broadcasters to be biased in their presentations, the multitude of congressional, state, and local candidacies calls for criteria to ensure equity. **As a first step toward total repeal of Section 315, we support elimination of the equal-time provision only for candidates for president and vice-president. Meanwhile, Congress and the Federal Communications Commission should develop and periodically review criteria by which broadcasters might allocate free time to candidates for congressional, state, and local office with a view toward complete repeal of the equal-time requirement when the increase in the number of channels indicates that it is no longer warranted.** *

THE PRESIDENT, CONGRESS, AND THE AIRWAVES

The President of the United States has almost always had unrestricted access to the broadcast media. It has generally been network practice to grant air time to the President whenever he requests it and in any time slot he may choose. Presidential ability to dominate the airwaves in this fashion has sharpened the mounting debate over the possible abuse of presidential powers. Unlimited presidential access to the media, it is argued, strongly shifts the balance of power among the three branches of government in favor of the President.

The FCC has consistently rejected the right of reply to a President.[3]

---

[3] Only once has the FCC ordered free time to be granted for a reply to presidential broadcasts. In 1970, in response to complaints from many sectors, the FCC studied network coverage of the Vietnam War and concluded that programming was generally balanced between pro- and anti-administration views, except for a great many presidential addresses. To correct this, the commission ordered networks to select a spokesman to present an uninterrupted program of opposing views. The commission, however, went to great lengths to explain that this was not a precedent for right of reply to a presidential address.

See memorandum by *C. WREDE PETERSMEYER, page 94.

Moreover, there is no law, policy, or practice that gives the legislative or judicial branches of government access to television or radio equal to that enjoyed by the executive branch. The fairness doctrine extends only to issues, not to individuals, organizations, or institutions.

Until recently, neither the Supreme Court nor Congress had sought equal access to the broadcast media. In fact, the Supreme Court has felt that it has been strengthened by lack of such coverage. Although the performance of the courts is of general concern, the precise issue before them at any given moment normally involves a dispute between parties. What takes place in Congress is the business of us all.

During the administrations of Presidents Johnson and Nixon, critics in Congress voiced increasing objections to the advantages that the President enjoys over Congress and the opposition party as a result of his unlimited access to the airwaves. As a means of offsetting presidential access to broadcasting and thereby checking and balancing presidential power, proposals have been advanced that would require networks to offer time for response to presidential broadcasts. Other proposals would mandate the airing of full sessions of Congress on a regular basis. A few Senate committee hearings have been broadcast over the years, and House committee proceedings have been opened for this purpose on a limited basis recently, but neither chamber has allowed its official debates to be broadcast by television or radio.

It is our view that the fairness doctrine affords reasonable opportunity for a balanced discussion of the many controversial issues that a President may address and that no equal-time or fixed-formula requirement imposed by the government is necessary to assure access for opposing views. Indeed, television and radio have often voluntarily provided opportunity for direct replies to presidential broadcasts and for presentations of opposing viewpoints when it was considered necessary for journalistic balance. Nor have the broadcast media failed to give day-to-day news coverage to the major activities of Congress.

At the same time, we have long supported measures to heighten the effectiveness of the legislative branch,[4] and we believe that the broadcast media, if allowed freer access to congressional floor debates and committee hearings, can help to achieve that goal. The 535 individual members of Congress act in their highest institutional capacities only when the houses of Congress are in full session. The business discussed in these

---

[4] *Making Congress More Effective* (1970).

sessions is the public's business and deserves public scrutiny. Broadcasters should be allowed to use their technology to bring Congress into full view.

We recognize that there is a risk television may create a circus atmosphere, with the participants playing to the cameras and the unseen audience. Conceivably, this could disrupt the legislative process. But these risks can be minimized by cooperation between congressional leaders and broadcast officials in the unobtrusive placement of equipment and in careful attention to production details. Broadcasts of the impeachment deliberations of the House Judiciary Committee in 1974 did not interfere with the proceedings. In fact, they were a significant public service. The greater risk is in depriving millions of television viewers and radio listeners of an opportunity to examine at close range the important work of their elected officials.

Broadcasters should be allowed to transmit by television and radio important events taking place on the floors of the Senate and the House and in committee hearings, subject to rules established by Congress in consultation with representatives of the broadcasting industry.

One of the most important ways of implementing this proposal would be to broadcast congressional debate on the federal budget. Under the Congressional Budget and Impoundment Control Act of 1974, Congress will for the first time review the federal budget as a whole and recommend overall levels of spending, revenue, and debt, just as the President now does in his budget message. A prerequisite for developing such a congressional budget (or, indeed, any federal budget) is full and open public discussion of the national priorities and choices that a federal budget implies. Congress is the most effective forum for focusing public attention on critical budgetary alternatives, and the broadcast media can play a crucial role in bringing these alternatives to public view.

SOCIAL PERFORMANCE OF BROADCASTERS

The public's impression of commercial television has shifted considerably over the years. In the early 1960s, public attention was focused on television primarily as an entertainment medium. Today, the emphasis is on news and information. Although entertainment continues to fill the bulk of the broadcasting schedule, the public now sees television as its major—and most credible—source of news and information. Moreover, according to a study prepared for CBS by the Bureau of Social Science Research, "the new focus on the news and information content of television has undoubtedly altered people's views about other various aspects of the medium's role—from how it affects the twelve-year-old to whether it is a benign or malevolent force in society."[5]

The explanation for this change in public awareness lies at least partially in the role television has played in bringing to the home screen the coverage of events of the most profound social and political significance: the assassination of President Kennedy, man's conquest of space and his landing on the moon, the urban riots, massive civil rights demonstrations, battlefield reports on the war in Vietnam, the fall from power and resignation of a President and the inauguration of his successor.

Television was an active participant in these events, not a passive bystander. It added to their impact and in certain cases helped to shape their outcome. In doing so, television has made a deep imprint on American society, and public expectations of the medium have soared as a result. A growing proportion of the viewing public now wants television to show an even greater awareness of national priorities, values, and ideals.

As a consequence, there is mounting dissatisfaction among important groups in American society over the tone and content of certain television programs. Special concern is focused on programs portraying violence and sex; on children's programs, particularly the amount and quality of advertising they contain; and on a wide range of programs dealing with subjects that are believed to be improperly or incompletely treated. *

Recently, as public concern over recession and inflation, the energy crisis, and other economic problems has risen, television has faced a new challenge. Although its coverage of economic issues has increased substantially, it can no longer be content with merely devoting more time to these problems; it must explore their underlying causes. Commercial television has yet to develop the necessary expertise to treat complex economic subjects with depth and perception while still observing the unique demands and limitations of the medium.

In the policy statement *Social Responsibilities of Business Corporations* (1971), this Committee explored the conflict between a corpora-

---

[5] Robert T. Bower, *Television and the Public* (New York: Holt, Rinehart and Winston, 1973), p. 186.

See memorandum by *C. WREDE PETERSMEYER, page 102.

tion's maximum profitability and its debt to society. We believe that the social responsibility of the commercial broadcaster is greater than that of the average business executive because the broadcaster is bound not only by his social conscience but also by law (as a public trustee) to place the public interest above his own.

In the end, corporate social responsibility is a matter of enlightened self-interest. For the broadcaster, this concept embodies the view that a radio or television station, like any other business, can thrive only in a healthy society. But it also involves the combined threat of legal action and public pressures for increasing government intervention and regulation to force broadcasters to do what they are reluctant or unable to do voluntarily.

The regulations governing broadcasting place the responsibility for programming decisions squarely with the licensee. We believe that the licensee must retain full control over those decisions. Nonetheless, when management fails to carry out this responsibility, there are means by which the public can seek redress. For example, citizens and citizen groups can collect and analyze program data, monitor programs for potentially harmful effects, and report problems and shortcomings to station management and to the community. Violations of specific rules and regulations can be reported in either informal or formal complaints to the FCC or other appropriate government agencies. Interested parties can file petitions to deny license renewal or submit a competing application for the license.

TELEVISION AND VIOLENCE

Television broadcasters bear a particularly heavy burden of social responsibility for program content. Although still inconclusive, the evidence suggests that programs depicting various forms of brutality and bloodshed may become deeply etched in the public mind, particularly in the minds of children, and that such programs could lead to antisocial behavior. The way in which the broadcasting industry deals with this issue may well establish a pattern for resolving conflict and adjusting to changing social needs *voluntarily*, without the threat of censorship or more stringent government regulation.

In its 1972 report, the Surgeon General's Scientific Advisory Committee on Television and Social Behavior presented preliminary and tentative indications of a causal relationship between viewing violence

See memorandum by *C. WREDE PETERSMEYER, page 103.

and aggressive behavior.[6] However, there are also indications that this relationship holds only for some children who are already predisposed to be aggressive and only in certain environmental contexts.

Considerable research is still under way to quantify the depiction of violence on television and to evaluate its effects. There is no doubt that violence is displayed on the television screen. But does it have legitimate dramatic purpose, or is it merely exploited for maximum shock effect? And to what extent do violent crimes portrayed on television influence members of the audience who subsequently commit similar or identical crimes?

However incomplete and tentative the findings, the causal relationship between televised violence and antisocial behavior is sufficient to warrant remedial action. The broadcasting industry is being held accountable for televised violence by a considerable segment of the viewing public. The burden is on the industry to conduct further research and provide more information about the effects of television violence. Therefore, we commend the television networks for the steps they have already taken in commissioning objective research and for the efforts they are making to resolve the problem voluntarily.

The broadcasting industry should join with social scientists and qualified representatives of the public to devise techniques to measure and control the amounts and types of violence on television, to evaluate the impact of programs portraying violence, and to inform the public about the problem as a means of protecting children against possible excesses. To accomplish this, the industry, along with private organizations and government agencies, should continue to support objective research and should be guided by its findings.

Suggestions of ways to limit or control the amount of violence seen on television by children are complicated by the fact that many parents are either unwilling or unable to supervise the viewing habits of their youngsters. No attempt should be made to censor programming or to homogenize programming to suit all tastes. But a method of alerting parents to a program's suitability for children that is acceptable both to the industry and to the public could go a long way toward solving the problem of violence on television.

---

[6] U.S. Surgeon General's Scientific Advisory Committee on Television and Social Behavior, *Television and Social Behavior: Report* (Washington, D.C.: U.S. Government Printing Office, 1972).

Detailed federal regulation of content is clearly not indicated; it would only invite government involvement in broadcasters' programming decisions. But if broadcasters do not solve the problem themselves, restrictive regulation or legislation may well occur. A high sense of social responsibility requires attention and constructive action on this problem.

GOALS FOR BROADCASTING

The CED statement *Social Responsibilities of Business Corporations* made clear that businesses cannot solve all the problems of society but that they can formulate goals and guidelines which will help them to determine what their most appropriate and effective social role might be. Broadcasting is now the dominant means by which Americans acquire information about themselves, their society, and their political process. *Broadcasting has a responsibility to establish identifiable goals and objectives that can provide a measure of success or failure in serving the public interest.*

We do not intend to propose a set of goals or objectives for any sector of the broadcast industry, but we do believe that some start should be made in shaping a set of general principles which can help to define the position of commercial broadcasting in future communications policy. We recognize the enormous difficulties involved in the process of setting goals. First, there is little agreement among those concerned with commercial broadcasting (the broadcast industry, the government, the broadcast audience) regarding what constitutes appropriate standards of social performance. Second, goals for broadcasting must be flexible enough to serve the many and varied segments of the nationwide commercial broadcast community. If national goals originate with only one group, no matter how representative or well intentioned, the risk of censorship is bound to arise. Third, if goals are formulated by too many disparate groups, the results can be too diverse and fragmented to be effective.

In a series of policy statements,[7] this Committee has recommended that government agencies, health organizations, and colleges and universities establish goals and objectives so that the public can understand their

---

[7] *Improving Federal Program Performance* (1971), *Building a National Health-Care System* (1973), and *The Management and Financing of Colleges* (1973).

purpose and can judge them by their own criteria. We see no reason why this should not be done in the case of the broadcast media.

Among other things, goals for broadcasting should be a thoughtful expression of industry standards. One of the best efforts of this kind is the Television Code of the National Association of Broadcasters (NAB). (Excerpts from the code appear on page 42.) It is concerned with the content of programs and commercials and with the amount of time permitted for advertising. Explicit in the code are specific goals and objectives for broadcasters. A major shortcoming, however, is that although it is held up as a model of industry self-regulation, it is unenforceable and in certain areas, such as maximum allotted time for commercials or violence in entertainment, is often not adhered to by subscribers. Moreover, no subscriber has ever been deprived of association with the code for violating any of its principles.*

Although we urge stricter adherence to the code, we also believe that other voices should contribute to the process of establishing meaningful and workable goals for broadcasting. Networks, broadcast industry groups, and individual stations should all play a role in this process, but so should groups that represent the broadcast audience. High on their agenda should be issues such as broadcasting's political responsibilities, the picture it presents of contemporary society, the mix between entertainment and public affairs and between programming and commercials, and the presentation of varying points of view. If serious public discussion of these goals can begin, priorities will emerge, and we can start to discover what the most appropriate role for commercial over-the-air broadcasting is and what roles should be assumed by other modes and technologies in a national communications system.*

See memoranda by *C. WREDE PETERSMEYER, page 103.

PRINCIPLES GOVERNING PROGRAM CONTENT

(Excerpts from the Television Code
of the National Association of Broadcasters)

It is in the interest of television as a vital medium to encourage programs that are innovative, reflect a high degree of creative skill, deal with significant moral and social issues, and present challenging concepts and other subject matter that relate to the world in which the viewer lives.

Television programs should not only reflect the influence of the established institutions that shape our values and culture, but also expose the dynamics of social change which bear upon our lives.

To achieve these goals, television broadcasters should be conversant with the general and specific needs, interests, and aspirations of all the segments of the communities they serve. They should affirmatively seek out responsible representatives of all parts of their communities so that they may structure a broad range of programs that will inform, enlighten, and entertain the total audience.

Broadcasters should also develop programs directed toward advancing the cultural and educational aspects of their communities.

To assure that broadcasters have the freedom to program fully and responsibly, none of the provisions of this Code should be construed as preventing or impeding broadcast of the broad range of material necessary to help broadcasters fulfill their obligations to operate in the public interest.

The challenge of the broadcaster is to determine how suitably to present the complexities of human behavior. For television, this requires exceptional awareness of considerations peculiar to the medium.

Accordingly, in selecting program subjects and themes, great care must be exercised to be sure that treatment and presentation are made in good faith and not for the purpose of sensationalism or to shock or exploit the audience or appeal to prurient interests or morbid curiosity.

Source: Reprinted from the Television Code, published by the Code Authority, National Association of Broadcasters, Seventeenth Edition, Second Printing, January 1974.

"Sesame Street," created to stimulate the education of preschool children, appears on some 250 public broadcasting stations and a number of commercial outlets. "The growth of new technologies will place heavy demands on public stations to help meet the programming needs of cable, film, and cassette. If public broadcasting is to continue its quest for excellence, maintain its independence, and reach its full potential, it must have adequate funding on a continuing basis."

3. A Policy for Public Broadcasting

THE PROMISE OF GREATER QUALITY, diversity, and choice in television programming was publicly proclaimed in 1967 by the Carnegie Commission on Educational Television:

Through the diversified uses of television, Americans will know themselves, their communities, and their world in richer ways. . . . Public Television is capable of becoming the clearest expression of American diversity and of excellence within diversity.[1]

We believe that public television can fulfill this promise, but not without greater funding on a continuing basis. At the same time, we recognize that money alone does not guarantee excellence in programming. Public broadcasting must identify and address the needs of its audience. The success of any proposal for increased long-term financing will depend on the establishment of firm, realistic goals and objectives for the future of public broadcasting and the effective management of its resources in achieving those goals.

Public broadcasting must also determine how it will fit into a new era in which cable television and other technologies could substantially widen program diversity and choice. In the 1960s, public television's growth was explosive. The number of stations increased from a little more than 50 to over 200. The number of employees nearly tripled; annual operating budgets increased more than fivefold; broadcast hours, sixfold. But will the same kind of growth benefit public broadcasting in the next decade and beyond? In anticipation of a new abundance, public broadcasting may well have to redirect its growth toward stimulating new sources of programming and new and expanded means of delivering programs.

HOW PUBLIC BROADCASTING EVOLVED

Although noncommercial radio broadcasting had been in existence since the early 1920s, the major impetus for public broadcasting came in 1952 when Frieda B. Hennock, a farsighted FCC commissioner with strong backing in the educational community, fought for and won the

---

[1] Carnegie Commission on Educational Television, *Public Television, A Program for Action: Report and Recommendations* (New York: Harper & Row, 1967), p. 18.

reservation of television channels for educational stations. During the 1950s, public broadcasting (or educational broadcasting, as it was known then) grew slowly, supported by state and local taxes, private contributions, and foundation funding. However, such support was not enough. In 1962, after a year of debate, Congress passed the Educational Television Facilities Act, making available $32 million in matching grants for the construction of facilities.

The fortunes of educational television took a sharp upward turn in 1967, the year the Carnegie Commission on Educational Television released its report. One of the report's key contributions was its emphasis on diversity in programming. The commission envisioned a system of local stations that would serve the many specialized audiences not reached by commercial television. It also emphasized the need for an interconnecting system for the distribution of quality national programming. But perhaps the most visible change resulting from the report was the introduction of the term *public television*. The commission believed that noncommercial television could do more than instruct its viewers; it could serve the larger public good.

The recommendations in the Carnegie Commission report were the basis of the Public Broadcasting Act of 1967, which authorized a nonprofit, nongovernmental corporation to promote and help finance public television and radio stations. The Corporation for Public Broadcasting (CPB) was established to foster a strong and effective system for public broadcasting. To that end, it had responsibility for receiving and distributing federal and nonfederal funds, for the production and distribution of national programming, and for channeling money toward the development of local and regional programs. It was also charged with providing a national interconnection of noncommercial stations. In 1970, CPB joined with a group of representatives of noncommercial television and radio stations and created the Public Broadcasting Service (PBS) and National Public Radio (NPR). PBS selected, scheduled, and promoted programs for public television stations; NPR produced national programs for radio and distributed those and other programs to noncommercial stations.

The Carnegie Commission specifically ruled out a government broadcasting system similar to Great Britain's BBC or Japan's NHK on grounds that American public broadcasting should be an independent, indigenous system reflecting national traditions and responding to national needs. The idea that public television might become a national "fourth network" has surfaced at various times. But in 1974, CPB and the public television stations approved an agreement leaving program choice to the local stations, a principle we strongly endorse.

Under a station program cooperative plan that went into effect in late 1974, certain programs will be produced only if individual stations pool their funds to pay for them. Stations will be provided with funds from CPB and other sources; for every $4 the stations commit to specific programs, the national pool will contribute $5. Although individual stations have always been responsible for program choice, the purpose of this new arrangement is to have local stations gradually assume the entire responsibility for support of the cooperative and for CPB to turn its attention to the development of new programming. PBS, which is owned and operated by the local television stations, has become the national coordinator and distributor of programming under this plan.

LONG-RANGE FINANCING

Public television and radio are beset by spiraling costs. They face increasing audience sophistication, growing demands for special-interest programming, and advancing pressures to modernize through the use of color, better remote coverage, newer facilities, and improved signal quality. In spite of the rapid growth of the system, public broadcasting's revenues have not kept pace with its increased costs and responsibilities. When figures are adjusted for inflation, the average station is now worse off financially than it was before the Public Broadcasting Act was passed.

If public broadcasting is to continue its quest for excellence, maintain its independence, and reach its full potential, it must have adequate funding on a continuing basis. The report of the Carnegie Commission offers these goals for public broadcasting: "What we recommend is freedom. We seek freedom from the constraints, however necessary in their context, of commercial television. We seek . . . freedom from the pressures of inadequate funds. We seek . . . freedom to create, freedom to innovate, freedom to be heard in this most far-reaching medium."

For public broadcasting, freedom, creativity, and innovation are expensive commodities; unhappily, there has been little recognition of their importance and cost in the authorization of federal funds. In 1973, Congress passed, and the President signed, a two-year authorization amounting to $55 million in fiscal 1974 and $65 million in fiscal 1975, but these amounts do not even approximate public broadcasting's real needs. Until public broadcasting is freed from unreasonable financial constraints and given adequate, sustained financial support, it can never hope to achieve the goals set forth by the Carnegie Commission.

We believe that both local and national public radio and television activities require a steady, long-range program of financing at levels consistent with established needs. Support must continue to come from all sources: individuals, foundations, universities, corporations, and federal, state, and local governments. However, if the federal government is to remain a minority partner in the public broadcasting enterprise, as we believe it should, most of the support for public broadcasting should come from nonfederal sources.

In considering various plans for federal financing of public broadcasting, we have rejected the proposal for a dedicated tax on the sale of receiving sets and similar proposals for earmarking specific revenues from a single source because we believe that money for public broadcasting should be subject to the congressional appropriations process. Instead, we are calling for increased support from general federal tax revenues.

We believe that federal financing must be provided in a manner that insulates public broadcasting from political intrusion. Public television will never develop fully if it is subject to sudden changes in the political climate. This goal can be accomplished by making federal money available on a long-term matching basis in a manner that stimulates support from nonfederal sources. Moreover, long-term funding is essential for accommodating the long lead time required for the planning and production of quality programming. Once a matching formula has been decided, it will be possible for federal funding to be increased year by year without political interference.

In July 1974, the Office of Telecommunications Policy (OTP) sponsored a bill under which the government would contribute $1 for every $2.50 raised from nongovernment sources. The bill called for a maximum federal outlay of $70 million in fiscal 1976, up to $100 million in fiscal 1980, for a total of $435 million. In August 1974, the Senate Commerce Committee raised the appropriation to ceilings of $88 million in fiscal 1976 and $160 million in fiscal 1980, for a total of $612 million, or $177 million more than the original OTP legislation recommended. In February 1975, OTP introduced new public broadcasting legislation with funding levels identical to the 1974 bill.

Both proposals fall short of the findings of the CPB Task Force on the Long-Range Financing of Public Broadcasting, which recommended, "in the light of today's realities," federal appropriations of up to $100 million in fiscal 1975 and up to $200 million by 1979, for a total of $750 million. The task force also recommended matching $2 of nonfederal funds with $1 of federal funds.[2]

Although we endorse the concept of long-range financing contained in these proposals, we nevertheless feel that they are limited by artificially low ceilings based on what their sponsors deem to be politically acceptable amounts rather than on the real needs of the system. Public radio and television are vital and growing elements in the nation's communications system. The growth of new technologies will place heavy demands on public stations to help meet the programming needs of cable, film, and cassette. Public broadcasting should be funded at a substantially higher level than any of these proposals have suggested. Our recommendations for the long-range federal financing of public broadcasting are stated in the following paragraphs:

Support of public broadcasting from general federal tax revenues should be authorized and appropriated by Congress for a period of no less than five years. The level of federal support for public broadcasting in any fiscal year should match nonfederal support on a one-to-two basis up to an established ceiling based on realistic costs of providing expanding quality broadcasting service. Federal matching funds are a well-established means of providing support and offering a strong incentive to maintain and increase state, local, and private revenue. In fact, matching has been the traditional form of federal assistance for building and improving local public broadcasting facilities and for many other categorical-grant programs. The one-to-two formula ensures that public broadcasting will remain principally a *nongovernmental* enterprise, free from federal control.

We must stress, however, that some measure of accountability is needed if public broadcasting is to gain the large-scale federal support it is seeking. Until channel abundance makes it unnecessary, public broadcasting stations should be subject to the fairness doctrine, the equal-time requirement, and other regulations that are designed to ensure balance and equity in programming on commercial stations.

Once federal support is appropriated, it should not expire at the end of the fiscal year but should be available until expended, as is cur-

---

[2] Corporation for Public Broadcasting, *Report of the Task Force on the Long-Range Financing of Public Broadcasting* (Washington, D.C., 1973), pp. 5, 10.

rent practice. **The distribution of funds should be made by the Corporation for Public Broadcasting in consultation with representatives of the public broadcasting stations.** Allowing appropriated funds to be available beyond the fiscal year would assure private contributors that there would be federal matching funds based on their actual contribution. It would also help avoid the trap of wasteful end-of-year spending, encourage efficient management, and allow for flexibility in the use of money. We further believe that it is essential for local stations to have a voice in deciding the distribution of funds. The station cooperative plan has established an appropriate mechanism for this purpose.

Beyond the funds provided by the recommended matching plan, federal matching grants for broadcasting facilities should be continued by Congress at the present three-to-one, federal-to-nonfederal level. Although we have given priority to federal support for programming, both public television and public radio are still seriously handicapped by inadequate funding for facilities. Most public television stations still cannot produce and broadcast in color. Many are technically disadvantaged by lack of sufficient antenna height and by low transmitter power. Two-thirds of the public television stations are located in the UHF range, which can be a handicap in a largely VHF market. Increasing the number of VHF public television stations should be a major goal of the facilities-grant program.*

Both operating and facilities funds should be made available for disbursement at the beginning of each fiscal year in order to encourage wise planning and the orderly use of resources.

Local, state, and national public broadcasting units should encourage greater support from private sources. If federal money is to be provided on a matching basis, all sectors of the public broadcasting system must actively encourage support from businesses, foundations, and individuals. We believe that the business community has an especially important role to play as an underwriter of innovative programs, both national and local.

STRENGTHENING MANAGEMENT

The combined pressures of rising costs and limited income make it necessary for public broadcasting stations to use their resources carefully. Public broadcasting stations are not businesses in the ordinary

See memorandum by *C. WREDE PETERSMEYER, page 104.

sense, but they do share many of the management problems faced by business organizations: problems of analysis, planning, control, and accounting. In the use of their resources, public broadcasting stations, like other organizations, should be governed by goals, objectives, and priorities.

When measured in terms of capital investment, operating budgets, and numbers of employees, many public broadcasting stations are sizable in scope; yet, their internal management systems are often those of small, basic operations. Regardless of the size of the station, principles and techniques of sound management can increase productivity and effectiveness and heighten accountability to viewers and to the public.

The Corporation for Public Broadcasting, in collaboration with appropriate business and professional organizations, should provide local television and radio stations with a comprehensive program for improved management, including opportunities for management training and standardized models of budgeting and accounting procedures as well as guidelines for their local application. The implementation of improved management methods should be a major responsibility of local station managers and boards of trustees.

GAINING AN AUDIENCE

Although public television has the capacity to reach over 150 million people and public radio over 130 million people, audiences are, in general, relatively small. Ratings for most public television programs in most areas are so low that they fall within the statistical margin for error. Only in large metropolitan areas or for special, well-publicized programs or events of national importance does public television command a mass audience.

For commercial broadcasting, attracting a substantial audience is usually the determining factor in whether a program survives or fails; hence, audience research is a basic tool. Audiences for public broadcasting are harder to identify. Public broadcasting does not have an economic incentive to attract a mass audience, nor does it have simple criteria for success and failure.

We believe that public broadcasting should have a built-in system for evaluating the programming needs of its specialized audiences. Although this type of advance market research is costly and therefore not routinely conducted in public broadcasting, it was a vital element in program planning for public television's most spectacular success, "Sesame Street." The audiences for "Sesame Street" were carefully researched in advance. The originators of the program consulted such diverse experts as psychologists and illustrators of children's books. A team of educational researchers spent eighteen months studying the interests and reactions of children in all parts of the country. Because of this initial research, the "Sesame Street" producers clearly addressed the needs of their audience. The result is one of the most universally acclaimed and genuinely popular programs on public television.

We are not suggesting that public broadcasting compete for an audience on the same terms as commercial broadcasting. What we are suggesting is that public broadcasting make a major effort to assess the interests and needs of its audiences and to determine what types of programs will meet those interests and needs and what criteria should be used to evaluate program success. Too often, policy planners in public broadcasting frame their objectives in vague generalities such as "to provide an alternative to commercial broadcasting" or "to serve the unfulfilled needs of the viewing public." Although public broadcasting does occasionally attract a large general audience (as it did with the Watergate and impeachment hearings), it should not make high ratings its major goal. It should, however, continue to develop high-quality general-interest programming that might also draw new viewers and listeners to other, more specialized offerings. Public broadcasting must state its programming goals in specific terms: for example, "to prepare programs designed to raise the reading level of children in grades 3 and 4" or "to reach one-tenth of the estimated audience for chess programs" (or symphonic music, public affairs, and so forth). It is only by defining specific programming goals and objectives that public television can reach and serve its audience. And it is the only way that cost-effectiveness, which is increasingly the standard for government and nonprofit funding, can be measured.

We propose that PBS prepare an annual report on the special interests of audiences that might be served by public television. The report would combine public broadcasting audience surveys with a wide range of audience data and opinion research from outside sources. The cost of the first report would be high, but the periodic updating would be considerably less expensive, especially when balanced against the great advantages of having such detailed information available. Public television would be better able to program for identifiable audiences and thereby eliminate costly programming for which there is little discernible interest.

We recommend that management at all levels of the public broadcasting system develop principles and techniques for determining the interests and needs of its audiences, the kinds of programming that will reach those audiences, and the criteria to be used for evaluating a program's success. We urge the Public Broadcasting Service to prepare an annual report indicating the special interests of audiences that might be served by public television.

PERFORMANCE EVALUATION FOR PUBLIC TELEVISION

In commercial television, the audience ratings serve as both a measure of accountability and a measure of business success. A high rating may indicate that a program is reaching millions of viewers who are potential purchasers of a television advertiser's product or service. Public television need not measure its success by sheer numbers, but it should conduct performance evaluations to determine how successful its programs have been in achieving their stated goals with the audiences they seek to reach. For example, the producers of "Sesame Street" and "The Electric Company," its counterpart for older children, know that even if their programming material matches their stated goals, these programs cannot be considered successful unless they help children to read. To measure program success, Children's Television Workshop, the producing organization, works with such agencies as the Educational Testing Service. The service has developed a group of tests for children who watch the programs, measuring improvement on the basis of specific curriculum objectives. Children's Television Workshop also has its own staff of researchers who supply feedback from viewers and recommendations for improving various projects.

Although such massive feedback from all public programming is not possible and probably not necessary, what is necessary is an evaluation system to determine the success of individual programs in reaching their target audiences, whether national or local. Public television must be accountable to the audiences it is supposed to serve. Without some way of developing programs to serve a known audience, it can never hope to attract significant support. But with an evaluation system designed for its specific needs, public television will have hard audience data with which it can justify the investment of public money, even though the figures will be much lower than those for commercial television.

The initial cost of such an evaluation system will be high. But since the goal is not overnight judgment, performance evaluations could be conducted monthly, quarterly, or even less frequently.

We recommend that the Corporation for Public Broadcasting establish a performance-evaluation system in order to determine how successful programs have been in achieving their stated goals with the audiences they seek to reach.

PUBLIC BROADCASTING AND THE NEW TECHNOLOGIES

Even with a massive infusion of new money, public broadcasting faces the same constraint that all broadcast media face: the scarcity of channels. But with the growth of cable and other new technologies, new delivery systems and totally new concepts in programming are possible.

Cable could become a major force in communications, and the role of public television and radio could change radically with such a development. With virtually unlimited channel capacity, there could be unlimited opportunity for programming to satisfy all tastes and needs. We have already noted that many public television stations are assigned to the UHF band, where the coverage a station can achieve is less than in the VHF band. Cable can offer these stations a larger potential viewing audience and can put them on an equal basis, in terms of picture quality and ease of tuning, with the more powerful VHF stations. Public radio can also achieve widened coverage through cable radio services. As we indicate in Chapter 4, the electronic media could become more like print, with the capacity to handle as much programming of as many different types as the public is willing to support.

If public broadcasting is to grow with the new technologies, it must adapt to changing circumstances and respond to new opportunities. Public broadcasting stations must begin to direct their efforts toward programming for cable and must explore the means by which they can gain greater access to cable channels outside their own coverage area. Public radio and television produce programming for specialized audiences, and both face the problem of choosing one type of program over another for any particular time slot. With cable's potential for multiple channels, several programs could be shown simultaneously either by leasing additional channels or by selling or renting programming to cable operators.

Instruction was one of the earliest goals of educational broadcasting, and cable could provide new outlets for instructional programming from the preschool level through adult education. Universities and municipalities that operate public broadcasting stations might explore the possibility of owning cable systems and diverting part of their revenues to noncommercial and instructional programming operations.

To program for an abundance of channels and a multiplicity of uses, public broadcasters should consider expanding their broadcast facilities into local and regional public telecommunications centers, which would create and produce educational, instructional, and cultural programming or would provide facilities for others to do so. This material could be made available to public television and radio stations, commercial stations, cable systems, schools, and individual users (in cassettes).

The public telecommunications centers could develop libraries of video and audio materials and programs in film and cassette form, which could be distributed electronically or by mail. These branch libraries could supply materials to the Public Television Library, which now serves as a national depository of programs produced locally, regionally, or nationally for public television and which has already set up experimental video cassette viewing centers in several localities. The Public Television Library could more appropriately become a national public telecommunications center, housing a wide variety of audiovisual materials in forms compatible with the newer technologies. It would provide a valuable information resource for the nation and, as a clearinghouse, would help to eliminate the duplication of effort that now takes place in the production of public broadcasting programs.

The future relationship between technology and public broadcasting could move in many directions. Satellites, for example, could provide more economic distribution, the opportunity to distribute by function, and the ability for stations to join together to meet common functional needs. Public television should also continue work in captioning for the deaf, which allows those viewers with hearing problems to use special decoders to see captions with regular programming. Subsidiary Communications Authority (SCA) frequencies could provide special services for blind and other print-handicapped persons.

To assure a place for public broadcasting among the new technologies, public broadcasters should focus their efforts on programming for a wide variety of purposes: public television stations, commercial stations, cable systems, schools, and individual users. To accomplish this, we urge them to consider expanding their stations into local and regional public telecommunications centers. We also urge public broadcasters to plan now to adapt to other new technologies such as satellites and to offer special services for the deaf and the blind.

Cables strung on telephone or utility poles provide a conduit for cable television. "The basic product of cable is over-the-air commercial and public broadcasting programs whose signals are picked up from local and distant stations and transmitted with improved reception by cable systems to their subscribers. But cable is increasingly originating programs of its own or leasing channels to others for that purpose."

4. The Potential of Cable

MOST AMERICAN TELEVISION VIEWERS select programs from three, four, or five channels. In larger cities, the choice may be much greater. Cable offers the potential of widening the selection of channels far beyond anything presently available.* The capacity of the coaxial cable, unlike that of the broadcast spectrum, has no inherent limits. It can offer twenty, forty, or even eighty channels. Thus, cable offers new possibilities for educational, municipal, and cultural programs and (as we indicate in Chapter 3) new opportunities for public television and radio. Broadband technology not only improves reception of over-the-air broadcasts but also makes possible the "cablecasting" of specialized programs unavailable to broadcast audiences. In the more distant future, regional or national satellite interconnections of cable systems could distribute special-interest programs and a panoply of business and consumer services, such as banking and shopping.**

The term *cable* can be misleading, however, and policy discussions often ignore the fact that cable embraces three distinct forms of subscriber-supported communications service:

1. The original CATV (community antenna television) was a method of bringing improved television service to areas that had inadequate reception. In its early days, such a system typically carried from three to five channels.

2. A newer service strengthens existing signals, imports additional signals into areas already served by television, and in certain cases originates programs of its own or leases channels to those who do. This is a twelve-channel system (or in some larger cities a twenty-channel system) that provides more varied program fare. This service may involve direct subscriber payment on a per-program or per-channel basis (pay cable).

3. A projected forty- or eighty-channel system can provide not only a more diverse broadcast service with improved signal quality but also a wide array of nonbroadcast entertainment and information services, including two-way communications.

All three categories are similar in that they share a closed-circuit, multichannel delivery system. But only the first two form the cable television industry as it exists today. Cable today is mainly an extension of over-the-air broadcasting. It can pull in a signal that has previously been unavailable, or it can make an existing signal sharper and brighter. Cable *could* become a comprehensive communications medium in its own right, offering a wealth of entertainment and information services, but its performance has yet to match its potential.***

See memorandum by *C. WREDE PETERSMEYER, page 95.
See memorandum by **JOHN A. SCHNEIDER, page 104.
See memorandum by ***C. WREDE PETERSMEYER, page 105.

CABLE ECONOMICS AND REGULATION

The cable television industry is a classic capital-intensive industry. Its future development will be determined in great measure by the availability and cost of capital; yet, because of the economic and regulatory climate, venture capital is presently in very short supply. These difficulties are compounded by the fact that the construction of a cable system requires a heavy initial investment. Furthermore, the return in the early years is slow; it may be ten years or more before an investor realizes a substantial profit. For these reasons, cable is essentially a franchised monopoly in its service area.

It is clear that the prodigious amounts of private capital needed for cable construction cannot be obtained without the expectation of future profitability. The cable business derives most of its revenues from installa-

---

CABLE CONSTRUCTION COSTS

Costs of constructing a cable system have escalated rapidly over the years, and they continue to rise sharply. In 1964, the average cost per mile for a twelve-channel system was approximately $4,000. Today, $6,000 per mile is considered a minimum estimate for overhead cable, and local conditions can increase that cost many times. If cable can be laid in conjunction with utility lines, the saving could be several thousand dollars per mile; but if utility conduit space is unavailable, underground construction can inflate costs dramatically. In New York City, the complex underground conduit system, right-of-way problems, expensive labor, and other difficulties have boosted costs to more than $50,000 per mile. Adding further to these costs is a 1972 FCC ruling requiring the construction of twenty-channel systems in larger markets.

Sources: Martin H. Seiden, *Cable Television U.S.A.: An Analysis of Government Policy* (New York: Praeger Publishers, 1972), p. 40. Current estimates from National Cable Television Association and Sterling Manhattan Cable TV, Inc.

tion charges and regular monthly fees paid by its subscribers; a small proportion of its revenues is also derived from advertising and pay cable. Installation fees are usually no more than $20; monthly rates range from $4 to $9. The financial success of a cable system (i.e., its ability to cover its costs and show a return on investment) depends on the percentage of subscribers it can enroll from among those households passed by the cable.

There are currently more than 3,000 operating cable systems throughout the United States, reaching some 10 million households, or about 14 percent of the nation's television homes (see Figure 1). Cable's economic future rests with the remaining 86 percent of television households. Most of this potential audience is situated in the 100 largest markets, which have easy access to a number of over-the-air television channels. In these markets, federal policy prior to 1972 virtually prevented cable systems from importing distant signals.[1] That policy was designed to protect the existing over-the-air television service from what was regarded as unfair cable competition,* and it dampened cable's growth in the nation's large urban areas. Presently, only 15 percent of all cable systems are in or within thirty-five miles of the 100 largest markets. Under the rules adopted in 1972, cable is freer to expand into major markets, but the amount of competition it can offer through the use of broadcast signals remains limited.

WHAT CAN CABLE DELIVER?

The basic product of cable is over-the-air commercial and public broadcasting programs whose signals are picked up from local and distant stations and transmitted with improved reception by cable systems to their subscribers. But cable is increasingly originating programs of its own or leasing channels to others for that purpose. Larger cable systems are required to make one channel available for municipal, educational, and public-access programs (which give citizens a chance to speak out on any issue).*(See Figure 2.) Although costs are presently prohibitive, the cable industry envisions two-way communications systems, facsimile reproduction of newspapers and mail, and a wide range of business, home, health, educational, and municipal services.

---

[1] See Appendix A, "A Brief Regulatory History of Cable Television."

See memoranda by *C. WREDE PETERSMEYER, page 106.

The product of cable is often thought of as special-interest programs, but at present, the major portion of its product is movies, sports, and entertainment programs that appeal to relatively large audiences and compete directly with over-the-air television. Essentially, cable is equipped to provide as wide a variety of programs as subscriber fees, advertising, and direct subscriber payment are capable of supporting.

The cable industry justifies carving out a piece of television's mass audience on grounds that these markets can provide the revenues that will enable it to offer specialized programs and a diversity of innovative communication services. The key policy question is: To what extent should cable be allowed to compete with broadcast television for programs and audiences?

NATIONAL CABLE POLICY

Cable must first succeed as a business before it can realize its full potential as an innovative technology. It must deal constructively with the problems it faces as a young industry in times of economic stress. Its costs have risen beyond expectations, and potential subscribers, expecting more from cable than an auxiliary broadcast service, have become disillusioned. As a result, profitability has declined, investors have become disenchanted, and venture capital has become scarce.*

But cable's commercial success has also been limited by an uncertain regulatory climate. A national cable policy is urgently needed to establish a regulatory framework within which cable can operate with some assurance about the future. We are not suggesting that cable policy can or should be firmly fixed for all time. In fact, such an approach might well doom cable's future. Clearly, cable's regulatory posture must be flexible in the immediate years ahead. But the need for flexibility should not become an excuse for failure to formulate a national cable policy.

We believe that if cable is to become a significant means of widening the range of programming and information services available to the American consumer, it should be allowed to prove its value in the marketplace. It is not at all clear, however, that additional channels of communication per se necessarily mean a broader or better range of services. There are cogent arguments for protecting the established broadcast service if a competitive system deprives the public of present benefits without offering the prospect of future improvements. *What is needed, therefore, is a national policy that strikes a reasonable balance between*

See memorandum by •HERMAN L. WEISS, page 107.

TABLE 1. GROWTH OF THE CABLE INDUSTRY IN THE U.S.

Year	Total Homes (thousands)	TV Homes (thousands)	CATV Systems	Subscriber CATV Homes (thousands)	CATV Saturation of TV Homes (percent)
1952	44,760	15,300	70	14	0.1
1953	45,640	20,400	150	30	0.2
1954	46,660	26,000	300	65	0.3
1955	47,620	30,700	400	150	0.5
1956	48,600	34,900	450	300	0.9
1957	49,500	38,900	500	350	0.9
1958	50,370	41,925	525	450	1.1
1959	51,150	43,950	560	550	1.3
1960	52,500	45,750	640	650	1.4
1961	53,170	47,200	700	725	1.5
1962	54,300	48,855	800	850	1.7
1963	55,100	50,300	1,000	950	1.9
1964	55,900	51,600	1,200	1,085	2.1
1965	56,900	52,700	1,325	1,275	2.4
1966	57,900	53,850	1,570	1,575	2.9
1967	58,900	55,130	1,770	2,100	3.8
1968	59,900	56,670	2,000	2,800	4.9
1969	61,300	58,250	2,260	3,600	6.2
1970	62,700	59,700	2,490	4,500	7.5
1971	64,500	61,600	2,639	5,300	8.6
1972	66,280	63,500	2,841	6,000	9.4
1973	68,330	65,600	2,991	7,300	11.1
1974	69,859	69,400	3,158	8,700	12.5
1975	Figures unavailable	70,837	3,200	10,000	14.1

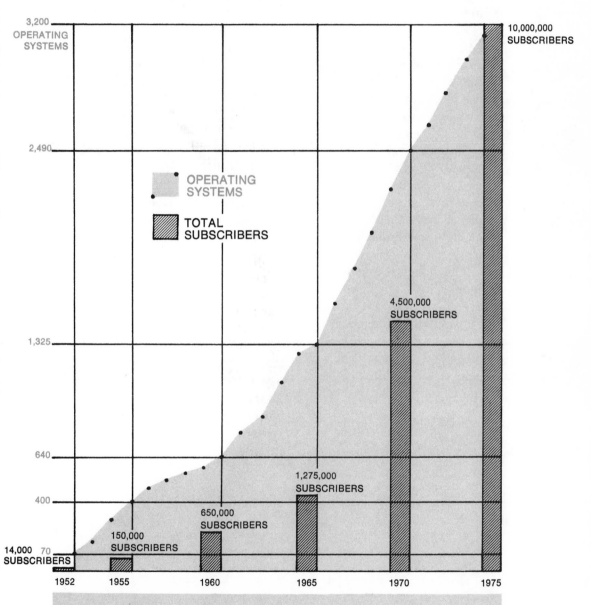

FIGURE 1. GROWTH OF THE CABLE INDUSTRY IN THE U.S.
(AS OF JANUARY OF EACH YEAR)

Sources: Figures for 1952 to 1974: *Television Factbook*, No. 44 (1974–75 ed.), services vol., p. 71–a. Figures for 1975: National Cable Television Association.

the promotion of diversity through cable and the preservation of an effective system of over-the-air broadcasting. Indeed, if a national cable policy is to generate the necessary political support, such a balance of interests is essential.

Our recommended policies for cable reflect a substantial consensus of this Committee. But it is fair to point out that widely conflicting viewpoints were expressed by a minority of the members. On one side were cable proponents who maintained that both law and regulation grant cable access to the broadcast spectrum and that cable should therefore be allowed to compete without regulation for audiences and for programs of all kinds. On the other side were advocates of over-the-air broadcasting who maintained that cable has built an industry "on the backs" of broadcasting by using its programs. They argued that the importation of distant broadcasting signals and the siphoning of mass-audience programs by cable constitute unfair competition which could stunt the growth and profitability of over-the-air television and perhaps destroy free television for the American consumer.

Our policy recommendations seek to resolve these divergent views by achieving a fair balance between measures that will allow cable to prove its worth in the marketplace and measures that will safeguard the great strengths of over-the-air broadcasting.

Before there can be agreement on the key elements of a national cable policy, there must be a clear concept of the goals that such a policy should seek to attain. We believe that a national cable policy should include measures that will achieve the following goals:*

Reasonable and regulated competition between cable and over-the-air broadcasting

Nondiscriminatory access to cable channels for program suppliers and users

Increased capital investment in cable and greater quality, diversity, and convenience in program presentation

Protection for program producers and creative talent against unauthorized use of their works when retransmitted by cable

There is no simple strategy for reaching these goals. No single measure will automatically produce an abundance of programs for the American consumer, nor will it provide the perfect competitive market for doing so. Rather, a variety of interrelated actions must be taken if cable is to offer the diversity of communications services that we believe is in the public interest.

See memorandum by *C. WREDE PETERSMEYER, page 107.

FIGURE 2. A CABLE TELEVISION SYSTEM

Tulsa Cable Television, Tulsa, Oklahoma, is a thirty-six-channel system. It carries the signals of the three local network-affiliated commercial television stations and one local public television station. It imports signals of independent television stations from Dallas and Fort Worth and offers a variety of programs on other channels. Twelve channels are not yet programmed.

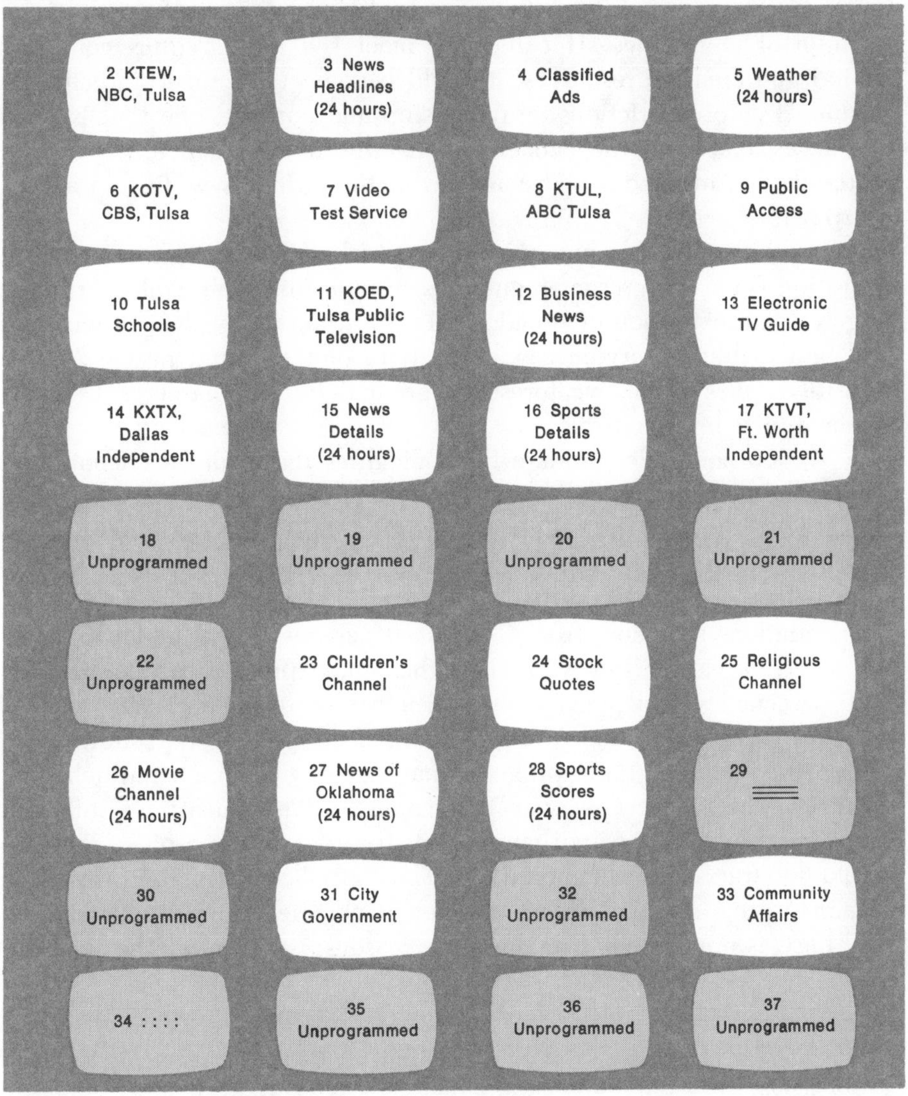

2 KTEW, NBC, Tulsa	3 News Headlines (24 hours)	4 Classified Ads	5 Weather (24 hours)
6 KOTV, CBS, Tulsa	7 Video Test Service	8 KTUL, ABC Tulsa	9 Public Access
10 Tulsa Schools	11 KOED, Tulsa Public Television	12 Business News (24 hours)	13 Electronic TV Guide
14 KXTX, Dallas Independent	15 News Details (24 hours)	16 Sports Details (24 hours)	17 KTVT, Ft. Worth Independent
18 Unprogrammed	19 Unprogrammed	20 Unprogrammed	21 Unprogrammed
22 Unprogrammed	23 Children's Channel	24 Stock Quotes	25 Religious Channel
26 Movie Channel (24 hours)	27 News of Oklahoma (24 hours)	28 Sports Scores (24 hours)	29 ≡
30 Unprogrammed	31 City Government	32 Unprogrammed	33 Community Affairs
34 : : : :	35 Unprogrammed	36 Unprogrammed	37 Unprogrammed

≡ Reserved for Convention and Tourist Channel : : : : Reserved for Tulsa Colleges

Cable is presently regulated as an adjunct to broadcast television, a medium whose capacity is limited by the scarcity of frequencies. Although the FCC's authority over cable is far from clear, cable remains subject to FCC regulation that addresses both the ownership and the programming concerns spelled out in the Communications Act of 1934.* As such, it is subject not only to ownership restrictions but also to the equal-time provision of the Communications Act, the fairness doctrine, and rules mandating the reservation of channels for special public purposes. Long-range public policy must be based on a view of cable as a medium of abundance. If cable can meet the large requirements for capital and creative resources, it could become (like print and film) a medium free of restrictions on ownership and content, one that is open and accessible, with the capacity to handle an abundance of program material of as many different varieties as the public is willing to accept and pay for.**

But a sound cable policy must also include measures to promote a transition from the present regulatory system, in which cable is treated largely as an extension of broadcasting, to a market system, in which it will be capable of serving a wider variety of tastes and needs. As this transition takes place, we foresee a gradual but ultimately substantial removal of federal controls.

We do not underestimate the difficulties that will be encountered during the period in which a gradual reduction of regulation must take place. The report of the Cabinet Committee on Cable Communications, which called for separation of ownership and program control as a long-term policy, suggested postponing this policy and other long-term recommendations that flow from it until 50 percent of the nation's homes are wired for cable.[2] But we believe that under present regulation, such a 50 percent level of penetration will *never* be reached.

The cable industry's overwhelming need in the short term is to obtain the necessary new program material to fill cable's multichannel capacity. This is the only way that cable owners can attract sufficient numbers of subscribers in urban markets to generate the profits necessary to obtain large-scale risk capital.*** Thus, a policy that allows the cable owner to program a minimum number of channels on a nondiscriminatory basis is imperative. For the same reasons, we support the selective

---

[2] U.S. Cabinet Committee on Cable Communications, *Cable: Report to the President* (Washington, D.C.: U.S. Government Printing Office, 1974).

See memorandum by *C. WREDE PETERSMEYER, page 108.
See memorandum by **CHARLES P. BOWEN, JR., page 108.
See memoranda by ***ROBERT R. NATHAN, page 92, and by C. WREDE PETERSMEYER, page 95.

relaxation of restrictions on ownership and a gradual phasing out of certain restrictions on pay-cable operations. A strengthened copyright law for protecting program owners and creative artists whose works are channeled to cable subscribers is badly needed and long overdue. Measures should also be taken to clarify the conflicting and overlapping federal, state, and local jurisdictions that cloud the regulatory environment in which cable operates.

The implementation of these policies should be carefully monitored to assure that they achieve the dual objectives of promoting program diversity on cable and preserving an effective system of over-the-air broadcasting.

If the past is any guide, these policies will not unfold automatically according to a precise schedule but will be shaped by the push and pull of political decision making. Thus, policy making for cable will not be an orderly process. However, our goal is to find a way to cope with the uncertainty and disorder that now exist, not to achieve perfect order.

A BALANCED STRATEGY

The following paragraphs examine the key elements in a national strategy for cable that will achieve the objectives of promoting greater program diversity through cable and of still maintaining an effective system of over-the-air broadcasting.

Our recommendations for cable are very closely interrelated. Any single recommendation followed in the absence of the others is unlikely to accomplish its stated goal adequately. They must therefore be comprehended as an integrated program.

These recommendations represent the Committee's best judgment concerning the measures required to fashion a balanced and coherent cable policy. They are nevertheless made with full understanding that information about cable is fragmentary at best, often contradictory, and far from conclusive. Additional information is needed to evaluate the market potential of both present and prospective cable programs and services. More should be known about how changes in government regulation, cable technology, and capital markets affect the industry's ability to provide varied and specialized programs and services. Data are lacking about the consequences for program producers, cable subscribers, and the general public of offering programs and services in competition with over-the-air broadcasting and other established technologies.*

See memorandum by *CHARLES P. BOWEN, JR., page 108.

Cable Ownership and Access. To assure an appropriate level of over-the-air broadcasting and to provide incentives for new programming, the cable-system owner should control the channels used for retransmission of local and distant over-the-air broadcast signals and should be allowed to originate or control programs on a limited number of additional channels. The remaining channels, representing the majority of those potentially available, should be leased by the system owner on a nondiscriminatory basis to all who wish to offer programs and services to meet the individual needs, interests, and tastes of cable subscribers.

In the early stages of cable's growth, there will be minimal danger that the system owner who originates his own programs might engage in practices designed to discourage competition from other program suppliers. Now and for some years to come, the major challenge will be to secure enough program material to fill cable's multiple channels and thereby increase subscriber growth. But as cable systems mature, there will be increasing incentives for the owner to try to monopolize the system. Steps must then be taken to require cable owners who originate their own programs to demonstrate affirmatively that these programs do not restrict competitive access to the system. For example, franchising authorities might require cable operators to make available for lease to others one equivalent channel for each channel used by the cable owner for program origination or for retransmission of broadcast signals.

Cable systems should be governed by a policy of nondiscriminatory access. Owners should be allowed to originate or control programming on a limited number of channels but should be required to demonstrate affirmatively that they are not restricting the competitive access of others.

If the cable industry is to receive the large infusion of capital, creative, and technical resources that it needs, federal regulations should not discourage broadcasters, publishers, or any other group, particularly groups that tend to have a large supply of those resources, from owning or investing in cable systems.

We believe that current restrictions on cable ownership are an inhibiting influence on the full development of cable and that they contribute to uncertainties about its future. Present FCC regulations ban ownership of cable systems by broadcast television stations within the community they already serve; ownership by television networks is banned altogether. Broadcasters and publishers are those most likely to be affected by competition from a successful cable system. They should not be prevented from owning or investing in cable systems or from supplying programs for cable in the markets they serve or elsewhere, providing this does not lead to single-owner domination of the media in a particular market.

We recognize the concern that media monopolies may limit the range of ideas and information available to the public and reduce competition for advertising and possibly for audiences. The federal government should be alert to the dangers of media concentration and should be prepared to take action to curb it through strict enforcement of the antitrust laws. However, exhaustive studies have found little evidence to justify such concerns, even in markets of relative media scarcity.[3] A second concern is that owners who are local broadcasters will not develop cable as aggressively as a nonlocal broadcaster would. However, we believe that cable's multichannel capacity, if governed by a policy that assures nondiscriminatory access for all channel users, considerably lessens the threat of media domination or anticompetitive practices resulting from common ownership of different media in the same market.

Under present economic conditions, it is unlikely that investment in cable systems would advance very far even if ownership restrictions were totally removed. However, this should not be a reason for maintaining ownership controls in markets in which they are clearly unwarranted.

Restrictions on the ownership of cable systems by broadcasters and networks should be relaxed to allow common ownership in selected markets in which a diversity of media and media owners already exists.[*]

Copyright for Cable. As early as 1931, Justice Louis D. Brandeis, writing for the Supreme Court, supported the notion of encouraging creativity in the electronic marketplace when he said that picking up radio signals and retransmitting them was a public performance for which the creators should be compensated.[4] In 1968, the United States Court of Appeals for the Second Circuit similarly held that the cable retransmission of distant television signals constituted a performance and was thus subject to a fee.[5] However, the Supreme Court ruled in 1974 that cable operators do not violate copyright law when they pick up and retransmit distant

---

[3] See Walter S. Baer et al., *Concentration of Mass Media Ownership: Assessing the State of Current Knowledge* (Santa Monica, Calif.: Rand Corporation, 1974).

[4] Buck v. Jewell-LaSalle Realty Co., 283 U.S. 191 (1931).

[5] Fortnightly Corp. v. United Artists Television, 392 U.S. 390 (1968).

See memoranda by *ROBERT R. NATHAN and by C. WREDE PETERSMEYER, page 96.

broadcast signals. Nevertheless, Associate Justice Potter Stewart, writing for the majority, made it plain that shifts in commercial relationships in the communications industry "simply cannot be controlled by means of litigation based on copyright legislation enacted more than half a century ago, when neither broadcast television nor CATV was conceived."[6] Major reforms in the copyright law are essential.

Cable is no longer merely an antenna system for over-the-air broadcasting; it is a commercial enterprise that charges its subscribers for a service and originates some programs of its own. Full applicability of the copyright law can provide important incentives for the creation of innovative programming for cable. It is urgent, therefore, that Congress enact new copyright legislation to protect the creative efforts of writers, composers, producers, and owners whose works are channeled to cable subscribers.

Determining just compensation for a creative product distributed by cable has been extremely difficult. More precise economic studies are required in order to fix reasonable fee scales. In the absence of such information, however, the parties concerned should jointly establish a schedule of royalty payments, and Congress could provide for compulsory arbitration if agreement between the parties cannot be reached.

Congress should modernize and strengthen the copyright law, making it applicable to the retransmission of programs picked up by cable systems from distant broadcast signals. *

Pay Cable. One of the most controversial issues in the development of cable involves the payment of a fee by subscribers above and beyond the basic monthly rate on a per-program or per-channel basis for programs that would not otherwise be available to them.

Proponents of the pay-cable format argue that the future of cable is closely tied to its ability to offer marketable programs for direct subscriber payment, particularly in major urban areas, but that present programming restrictions imposed by the FCC prevent cable operators from doing so. Hence, cable's major current concern is with removing these federal controls over the programming (chiefly movies and sports) that is crucial to attracting new subscribers in the bigger cities. Under present

---

[6] TelePrompter Corp. v. Columbia Broadcasting System, No. 72-1628 (1974); and Columbia Broadcasting System v. TelePrompter Corp., No. 72-1633 (1974).

See memoranda by *E. B. FITZGERALD and by C. WREDE PETERSMEYER, page 96.

rules, movies are available to pay-cable companies only during the first two years of their release; after that, they are not available again for ten years. Movies that are over ten years old may be shown at the rate of one a month for one week only. Continuing series featuring a fixed cast of characters are not allowed over pay cable. Certain sports events are prohibited unless they have not been carried regularly by commercial stations within the preceding two years.[7] *

Proponents view pay cable as a major incentive for continued investment in the cable industry and as the only means by which cable can fulfill its promise of program abundance and diversity. They claim that because pay-cable programs may be offered several times a day, every day of the week, present restrictions are depriving viewers of the right to watch programs in their own homes, uninterrupted by commercials, and at their own convenience. Beyond such mass-appeal programs as movies and sports, proponents promise a variety of new educational and cultural programs on a pay basis; this, they believe, can also provide new sources of income for the nation's financially troubled educational and performing arts institutions.

Opponents of pay cable, on the other hand, maintain that cable operators will eventually outbid advertiser-supported television for the best movies, sports, and other programming, thereby siphoning or drawing off programs that are now offered free, and that they will perhaps some day move all television from an advertiser-supported to a pay system. A total free-market system, without any controls, they contend, is not likely to add to the richness and diversity of the programming the public receives because new programming can be costly and difficult to develop. Rather, removal of all restrictions would simply divert the most popular programs from conventional over-the-air television to pay cable and thus make cable compulsory for many viewers who want to continue to see their favorite programs without charge. Those who stand to lose the most from relaxing present antisiphoning rules, they claim, are low-income

---

[7] On November 15, 1974, the FCC issued "tentative instructions" to its staff that would permit pay-cable operators to show new motion pictures within three years of their theatrical release, instead of the present two years. The proposed new rules would permit pay operators to offer both original series and any that had not played on commercial television during the preceding five years. The proposed rules would deny to pay cable certain sports events unless they have not been carried by commercial television for five years, instead of the present two years, but would loosen restrictions on games that television does not cover.

See memorandum by *CHARLES P. BOWEN, JR., page 108.

groups that cannot afford to pay for these programs and the residents of rural areas that cannot be economically served by cable.

We believe that a mixed system of support for cable programs should be encouraged. Using the print media as a model, cable should accommodate various forms of financing, including direct subscriber payment for programs and advertising, if it will increase the choice and quality of programs and offer them at different times that may better serve the convenience of the viewers.*

It is clear that substantially more information is needed before a definitive policy on pay cable can be developed. There is little empirical evidence either to support or to refute the contention that advertiser-supported broadcasting has been, or is likely to be, seriously threatened by cable. Moreover, pay-cable policies can affect different markets differently. Without substantial experimentation, the nature of demand for cable programs under various price, cost, and market conditions, as well as the consequences for cable subscribers and the general public, cannot be easily ascertained.

In the absence of such information, several policy options are available:

1. Freeze existing programming controls, and allow cable to evolve on the basis of its ability to improve signal quality and offer programs other than movies, sports, and series programs now available on advertiser-supported television.

2. Declare a limited moratorium (four or five years) on all pay-cable programming restrictions and allow pay cable to be tested freely in the marketplace. If evidence develops that pay cable is competing unfairly or harming the public interest, programming controls should be reimposed.

3. Phase out programming restrictions gradually and selectively, and authorize carefully controlled experiments designed to acquire empirical data on pay cable's impact on free over-the-air television service. If evidence develops that pay cable is competing unfairly or harming the public interest, steps should be taken to restrain it, perhaps including reimposition of the present controls.**

All three options offer both advantages and disadvantages, but we believe that the third option, that of phasing out controls gradually and selectively, is most consistent with the national cable policy goals set forth

See memorandum by *C. WREDE PETERSMEYER, page 108.
See memorandum by **EDWARD N. NEY, page 98.

earlier in this chapter. Only some 130,000 of the nation's 66 million television homes now subscribe to pay cable, so it poses little immediate threat to over-the-air television. It should be given a fair test so that evidence of its impact can be determined as experience with it is accumulated.

We are aware of the pitfalls in a gradual, experimental approach. On the one hand, if experimentation proceeds too slowly and is too selective, it will not give pay cable a fair test, and prudent financial management will not make the necessary private investment. On the other hand, if experimentation proceeds too quickly and is too extensive, it will be difficult to reverse, regardless of what the results may show. However, we do not advocate blind experimentation; we urge experimentation that is based on careful analysis and study and that offers high probability of success. Experiments should seek to determine whether pay cable erodes the audience for over-the-air television in significant numbers, whether it deprives the public of access to programs it can now watch on broadcast television, and whether it weakens the financial base of broadcast television to the point where it must curtail its present service. Experiments conducted in a variety of markets may shed light on a number of influences and consequences. To monitor experiments and review their results, the FCC should appoint an independent panel that would issue its findings and recommendations to Congress and to the public.

We see no alternative to experimentation despite its difficulties as a means of gaining more knowledge about the effects of pay cable on over-the-air television and its ability to serve the public.

Programming restrictions on motion pictures and series programs should be phased out gradually and selectively. The Federal Communications Commission should authorize and carefully monitor experiments designed to evaluate the impact of such deregulation on free over-the-air television service. If loosening program controls on movies and series programs leads to unfair competition or other developments injurious to the public interest, the Federal Communications Commission should take steps to curb such practices; these steps might include reimposition of the present controls.*

Sports events are a special case. We believe that certain restrictions on professional sports programs are justified on grounds that professional sports are particularly suited to over-the-air television presentation. Unlike movies, sports are not usually subject to editing, and they do not suffer, as other programs do, from commercial interruptions. Television and sports events have enjoyed an interdependent relationship that has produced a large and loyal audience, and the public's right to view the most popular sports events on television without direct charge should be protected. At the same time, however, only 30 percent of all professional sports events are now regularly televised. The pay-cable industry should be allowed to compete for games that are not shown regularly on over-the-air television.

Antisiphoning restrictions on major sports events should be maintained, but pay-cable regulations should be modified to allow the presentation of games that are not regularly televised. *

Two-Tier Regulation. The division of responsibility for cable regulation among federal, state, and local governments needs both rationalization and clarification. The present confusion and delay created by overlapping and fragmented jurisdictions will probably never be completely eliminated, but it can be substantially reduced through more orderly regulatory arrangements. We believe that a two-tier (federal and state) system of regulation can bring considerable order to the regulatory chaos that now exists.

The FCC should be responsible for establishing the jurisdictional framework of policies of state regulatory agencies but should not be allowed to preempt state powers. It should establish requirements concerning minimum channel capacity and minimum technical standards for the construction and operation of cable systems. Continuing federal supervision will also be required with respect to the authorization of carriage of over-the-air television signals and for the enforcement of federal regulations governing copyright and ownership and of laws governing libel, obscenity, and profanity. The federal government should authorize cable experiments and play a major role in furthering research and analysis in the cable field.

Each state should establish a special commission or agency empowered by legislation to assume major responsibility for the regulation of cable systems. The commission should identify appropriate franchise areas within the state and, where special circumstances dictate, delegate franchising powers to local governments, particularly large cities, and provide overall guidance to these governments for their franchising activities. It should establish standards for the allocation of leased channels on a nondiscriminatory basis.

The key justification for the reallocation of responsibilities from the local to the state level is the widespread lack of experience, technical capability, and management resources at the local level in many communities. This is not to say that many states are not similarly handicapped, but

See-memoranda by *CHARLES P. BOWEN, JR., by E. B. FITZGERALD, and by C. WREDE PETERSMEYER, pages 99 and 100.

they are better equipped than local governments to acquire the necessary resources and capabilities. Moreover, the states are in a good position to moderate boundary and jurisdictional disputes that inevitably arise among overlapping and fragmented localities in the consideration of cable franchises.

At present, only twelve states have assumed authority over cable.[8] Of these, only three have established special cable commissions for this purpose. The remainder have assigned this responsibility to public utilities or public-service commissions, whose primary concern is the regulation of common carriers and other utilities and businesses. These commissions are responsible for fixing the rate of return on investment, regulating the level and structure of rates, and enforcing the provision of uniform service. However, cable is both structurally and substantively different from other common carriers; therefore, its regulation requires a separate state agency that is attuned to its special characteristics and problems.

We recommend a two-tier system of government regulation of cable involving the federal government and the states. The Federal Communications Commission should establish the jurisdictional framework to assure that state regulation is consistent with national policies. State governments should establish state commissions with authority over cable-franchising activities and procedures. Where local conditions warrant, states should delegate franchising powers to local governments, particularly large cities with proven resources for the regulation of a cable system.

It is impossible within the scope of this statement to indicate all the various interim measures that will be required before our recommended policy for cable can take full effect. Nor can we plot with certainty the pace of change from present cable policies to those we are proposing. What is important, however, is the direction that these changes should take and the ultimate goals that we are seeking. National policies for cable should move toward a relaxation of federal controls and a more important regulatory role for state governments. They should be designed to promote diversity on cable while still maintaining an effective system of over-the-air broadcasting.

---

[8] See Appendix B.

Richard Wiley, chairman of the FCC, recognizes a speaker at a public meeting in Washington, D.C. "The FCC is the public's guardian of the airwaves. Although it is formally accountable to Congress, it is—or should be—ultimately accountable to the public and to the public interest."

5. Organizing for Change

THE CAPACITY OF GOVERNMENT ORGANIZATIONS to respond to change is vital to the advancement of public policies for broadcasting and emerging technologies. Government regulatory policy not only must deal with the immediate problems of the industries involved but also must anticipate the ways in which new technologies can affect the regulatory climate of the future. We believe that the organization and management of government activities in the field of communications must be modernized and strengthened. We give a high priority to measures that will remove judicial responsibilities from the FCC and place them with a communications court and that will bolster the research and analysis capabilities of the FCC with adequate funding and staff.*

FEDERAL COMMUNICATIONS COMMISSION

The FCC, like other regulatory agencies, operates in a complex environment in which many forces come into play. Its success in creating effective communications policy requires sensitivity to economic and technological conditions and to political power and social change. It depends on achieving the right blend of processes and techniques; legal, administrative, and technical resources; and political leadership.

The FCC assigns frequency bands to nongovernment communications services, licenses and regulates stations and operators, and regulates common carriers, including telephone, telegraph, specialized carrier, and satellite entities. It is administered by seven commissioners who are appointed by the President to serve for terms of seven years with the approval of the Senate. The commissioners are organized in a collegial body that supervises all FCC activities. They delegate many technical and administrative functions to staff units, but policy determinations are the responsibility of the commission as a whole.

One commissioner is designated chairman by the President. The chairman presides at all commission meetings, coordinates and organizes its work, represents it in legislative matters and in communications with other government departments, supervises all FCC activities, and delegates responsibilities to staff units, bureaus, and committees of commissioners.

The FCC has been a subject of study and scrutiny since its formation, and the overwhelming conclusion of this analysis is that the FCC

See memorandum by *JOHN A. SCHNEIDER, page 110.

has not been performing effectively. The general criticism of the commission is that it has had great difficulty formulating coherent, forward-looking communications policy.*

Robert E. Cushman, who advised the President's Committee on Administrative Management during the 1930s, declared in a report: "Neither the Radio Commission nor its successor, the Federal Communications Commission, has come to grips with the major policy problems which are involved in the regulation of the radio industry."[1] The Hoover Commission asserted that the FCC had "been far from successful in achieving continuity of policies in areas where it was possible to do so. Even during periods when the composition of the commission was fairly continuous, stated policies were either transgressed or neglected."[2]

In a 1960 report to President-elect Kennedy, James M. Landis commented: "The Federal Communications Commission presents a somewhat extraordinary spectacle. Despite considerable technical excellence on the part of its staff, the Commission has drifted, vacillated, and stalled in almost every major area. It seems incapable of policy planning, of disposing within a reasonable period of time the business before it, of fashioning procedures that are effective to deal with its problems."[3] President Johnson's Task Force on Communications Policy concluded that the agency did not have the "resources to develop sufficient in-house capability for the analysis of major issues having technical, economic, and regulatory dimensions, even when these issues are central to its regulatory responsibilities."[4]

These studies and others that this Committee has reviewed cover forty years of FCC activity and service by more than fifty commissioners

---

[1] Robert E. Cushman, *The Independent Regulatory Commissions* (New York: Octagon Books, 1972), p. 730.

[2] U.S., Congress, House, Committee on Interstate and Foreign Commerce, *Regulation of Broadcasting: Half Century of Government Regulation of Broadcasting and the Need for Further Legislative Action* (Washington, D.C.: Special Subcommittee on Legislative Oversight, 1958), p. 108.

[3] James M. Landis, *Report on Regulatory Agencies to the President-Elect,* (Washington, D.C.: U.S. Government Printing Office, 1960).

[4] U.S. President's Task Force on Communications Policy, *Final Report* (Washington, D.C.: U.S. Government Printing Office, 1968), chap. 9, p. 24.

See memorandum by *C. WREDE PETERSMEYER, page 100.

with widely divergent views and backgrounds. They provide a solid foundation on which to build a policy for reshaping the commission.[5]

A COMMUNICATIONS COURT

Major causes of the FCC's deficiencies are the increasing scope, complexity, and detail of broadcast regulation and the lack of time, staff, and budget to create and execute sound, forward-looking policy. The greatest portion of the commission's time is taken up with daily executive, administrative, and prosecutory matters. Lacking the time and wherewithal to develop long-range policies, it has had no alternative but to base its regulatory judgments on a history of past practices. Therefore, prior behavior is often incorporated into policy without reflecting social or technological change.

The FCC's reliance on traditional rules and precedents inevitably places any challenger to the established industry in the position of threatening a rigid, firmly entrenched system. Thus, the FCC has often been called a captive of the industries it regulates, isolated from the public interest and alienated from the broader aspects of public policy.

We believe that the trend toward the separation of rule making and adjudication in the FCC has helped to ease the commissioners' burden of settling individual cases and has allowed them to devote more time to establishing meaningful rules and national policy. But a complete separation of the functions of rule making and adjudication is in order. What is needed is a communications court that would assume the commission's present adjudicatory responsibilities. The establishment of a communications court would also free the commission from conflicts that inevitably arise when it must enforce the rules it has instituted.

---

[5] Other sources that provided background for the Committee's recommendations are: Leonard H. Marks, J. Roger Wollenberg, and Edward P. Morgan, "Revision of Structure and Functions of the Federal Communications Commission," *Federal Communications Bar Journal* 18 (1963): 4; Newton N. Minow, *Equal Time: The Private Broadcaster and the Public Interest* (New York: Atheneum Publishers, 1964); U.S. President's Advisory Council on Executive Organization, *A New Regulatory Framework: Report on Selected Independent Regulatory Agencies* (Washington, D.C.: U.S. Government Printing Office, 1971); and Henry Geller, *A Modest Proposal to Reform the Federal Communications Commission* (Santa Monica, Calif.: Rand Corporation, 1974).

The idea of a communications court was discussed by the 1955 Hoover Commission and more recently in a report by the Federal Communications Bar Association.[6] This latter report pointed out that because the FCC lacks the time, the professional background, and perhaps the inclination, it has decided its cases according to the institutional approach, which places study of the record and opinion writing with a staff that necessarily rationalizes the results previously arrived at by the commissioners. The report also noted that the FCC has never established adequate standards or criteria for reaching its quasi-judicial decisions.

We recommend that the adjudicatory functions now exercised by the commission be conferred upon a new communications court. This new court would follow the pattern of the United States Tax Court while it was part of the executive branch. The judges would be appointed by the President with the advice and consent of the Senate for terms of at least twelve years in order to assure their independence. They would sit individually, but a review by the entire court might be provided in matters of unusual importance. Appeal would be to the United States Court of Appeals for the District of Columbia Circuit, as is now the procedure with most FCC cases. The communications court should be financed by congressional appropriation, but the funding should be separate from that of the FCC. As a court of special expertise, it could act on an informed basis in all areas, even in the sometimes subjective field of comparative licensing.*

POLICY RESEARCH AND ANALYSIS

Research and analysis, conducted by the FCC and outside it, should play a larger role in determining national communications policy. Although we believe that government regulation of broadcasting will be necessary for many years, we also believe that regulatory policies must be flexible enough to adapt to developing technologies. A proliferation of diverse electronic media should bring about a corresponding decrease in the need for FCC regulation.

In order to prepare for this future abundance and diversity, the FCC should substantially strengthen its research and analysis capabilities. The

---

[6] Marks, Wollenberg, and Morgan, "Revision of Structure and Functions of the Federal Communications Commission."

See memorandum by *W. D. EBERLE, page 111.

commission has already recognized this need. In 1973, it established the Office of Plans and Policy, but the work of this office has only recently begun. The office was created to recommend research blueprints and projects to the commission and to evaluate and analyze proposals made by other offices and bureaus. It has a small staff of economists, engineers, attorneys, and other professionals and is charged with working with individual bureaus in originating proposals, coordinating various recommendations, and offering alternative courses of action for present and future policy.

We support the goals of this office and encourage Congress and the FCC to allocate sufficient funds for their accomplishment. The office should serve as a center for improving interdisciplinary research and analysis capacities both for the commission as a whole and for the individual bureaus. It should also serve the important function of coordinating policies for the interrelated policy aspects of the new technologies.

To strengthen its research and analysis capabilities, we recommend that Congress and the Federal Communications Commission give high priority to the growth and development of the Office of Plans and Policy. Funding should be sufficient to allow experienced economists, engineers, attorneys, and social and political scientists to provide strong policy research and analysis both for the commission as a whole and for the individual bureaus.

FEDERAL COMMUNICATIONS COMMISSION AND CONGRESS

The FCC is both an independent regulatory agency and an arm of Congress, subject to congressional funding and review. The failure of Congress to provide long-range guidelines and standards for the FCC has weakened the commission and made it more vulnerable to other forms of congressional and political influence. The President's Advisory Council on Executive Organization (Ash Council)[7] concluded that congressional participation in agency regulation is more necessary than ever "because of the increasing interdependence of national economic policies which

---

[7] U.S. President's Advisory Council on Executive Organization, *A New Regulatory Framework: Report on Selected Independent Regulatory Agencies* (Washington, D.C.: U.S. Government Printing Office, 1971), p. 15. Roy L. Ash was chairman of the council.

emerge from budget and fiscal action, economic regulation, and industry promotion by government."

In addition to neglecting its responsibility to formulate guiding policy, Congress, through the Senate, has yielded to considerable pressure from the executive branch and the industry in the appointment of FCC commissioners. Even though the Senate has the final authority over any FCC nominee, no attempt has been made to set definite standards for the President to follow in nominating new commissioners.

Another major congressional failing has been in the vital field of administrative oversight. The relationship between a legislative committee and the agency it oversees calls for a continuous feedback to the committee about progress, achievements, and problems, supplemented by regular annual hearings and quadrennial reviews. Although Congress itself may lack the information and expertise it needs to review the activities of the FCC, we believe it should make greater use of outside resources and objective, nonpartisan advice. Thus, it can better establish meaningful standards of FCC performance and better assure that the public interest is reflected in regulatory policy.

PLURALISM IN THE REGULATORY PROCESS

The FCC is the public's guardian of the airwaves. Although it is formally accountable to Congress, it is—or should be—ultimately accountable to the public and to the public interest. We feel, therefore, that the FCC must be especially sensitive to the views of the many groups that make up the public and that it should consider their interests in guiding the nation's communications system.

Business, ethnic groups, labor, consumers, government, and other special-interest groups all have a stake in broadcast policy. If the FCC is to fulfill its charge to act in the public interest, it must be constantly aware of the danger of overidentifying with the problems and interests of any single group. Such an awareness should lead to the determination to prevent any group or groups from exerting undue influence over its policy deliberations.

One mechanism that is often proposed to assure that the citizen's voice is taken into account in the making of broadcast policy is the establishment of a high-level citizens' advisory committee to observe the communications scene and issue findings and recommendations. However, we believe that yet another watchdog committee, yet another blue-ribbon panel, is not the answer. Such an organization could very easily become dominated by groups with power and influence, leaving the less powerful as disenfranchised as they were before.

It is in the FCC's interest to make its decisions in light of information available from many quarters, and it should seek out, hear, and consider as many interested and informed opinions as possible. In 1974, the FCC initiated a series of regional meetings at which the public can meet commissioners and staff members and air its problems. We strongly support such initiatives. Public meetings, both in Washington and around the country, can, if properly conducted, provide the FCC with a firmer grasp of national sentiment regarding the many issues under its jurisdiction.

What we are proposing is simply a more receptive attitude on the part of the FCC to a wide range of serious, well-considered views. *What is needed is a free flow of responsible influence, not from one group or a small cluster of groups, but from a great many sources.* Not every claim will be entitled to satisfaction, but each is entitled to be heard, and no one group should become so dominant that it submerges the claims of all others.

ROLE OF THE EXECUTIVE IN COMMUNICATIONS POLICY

On February 9, 1970, President Nixon submitted to Congress a reorganization plan creating the Office of Telecommunications Policy (OTP). This action continued a trend toward the accumulation of administrative and technical expertise in the executive branch that began with the President's Communications Policy Board, which President Truman created to cope with the competition between government and nongovernment users of radio frequencies. President Truman was the first President to make a comprehensive examination of the nation's communications requirements. During the Eisenhower administration, the functions of the Office of Telecommunications Advisor to the President, established by President Truman in 1951, were shifted to the Office of Defense Mobilization (ODM) in the Defense Department. ODM was authorized to coordinate government activities in telecommunications and report to the National Security Council. A second Eisenhower directive merged ODM with the Federal Civil Defense Administration to form one office, the Office of Civil and Defense Mobilization (OCDM). In 1958, OCDM's director established a Special Advisory Committee on Telecommunications, which

See memorandum by •C. WREDE PETERSMEYER, page 112.

recommended an executive office to make policy and carry out the President's communications responsibilities.

Before assuming office in 1960, President Kennedy commissioned James Landis to examine the federal structure for the management of telecommunications. The Landis report emphasized the need for strong executive coordination of both domestic and international communications. In 1962, President Kennedy established the Office of Telecommunications Management (OTM) within the Office of Emergency Preparedness.

Although OTM was a small office with a small budget, it became the focal point for the executive branch's interest in communications policy. The search for a stronger mechanism continued into the Johnson administration under the President's Task Force on Communications Policy, which called for strengthening the total policy-making capability throughout the government, both in the FCC and in the executive branch. The task force report recommended a new executive office to assist the FCC in gathering and updating operational expertise and to provide the President with the latest problem-solving and forecasting techniques based on economic, technological, and communications systems analysis.

When President Nixon took office, much of the groundwork for centralizing communications policy had already been laid. The trend toward more centralized executive telecommunications control can be explained, in part, by the federal government's increasing need for sophisticated communications techniques. The government's total investment in telecommunications is more than $100 billion, and its annual expenditure for equipment, research, and development exceeds $7 billion.

OTP was established to serve as the President's principal advisor on telecommunications policy, to manage the government's own massive telecommunications system, and to prepare emergency communications capabilities. It was also charged with conducting and coordinating economic, technical, and systems analyses of telecommunications policies, activities, and opportunities and with developing (in cooperation with the FCC) a comprehensive long-range plan for improved management of all the resources of the electromagnetic spectrum.

OTP's main function is to advise the President. But in the early 1970s, it stirred controversy by encroaching on the regulatory authority of the FCC and issuing threats to both commercial and noncommercial broadcasters. This deviation from its stated purpose tended to obscure the valuable research and coordinating work being performed by OTP. There is a clear need in the White House for an advisory office that can explore the effective use of telecommunications technologies and concepts and can make policy recommendations to the President which can serve as a basis for recommendations to Congress for new or modified legislation. OTP must, of course, respect the division of powers among the executive branch, the FCC, Congress, and the courts. But as a White House advisory office, it should continue to examine the government's system of telecommunications and to provide communications expertise to the President. Moreover, OTP can provide a valuable service by making its advice, research, and technological expertise available to the FCC in a free, reciprocal exchange of information that would be valuable to both organizations.

Memoranda of Comment, Reservation, or Dissent

Page 12, by C. WREDE PETERSMEYER, *with which* MARVIN BOWER *has asked to be associated*

I do not approve of the statement and do not think it should be published. It proceeds from a series of assumptions and hypotheses that are themselves unsupported and yet build upon each other and permeate much of the statement. This approach makes a thorough recitation of reservations or dissents virtually impossible save in a document longer than the policy statement itself. Accordingly, this general comment and several specific notes will focus on some of the more basic problems presented by the policy statement, its methodology, inadequate supporting material, and some of its conclusions and recommendations. The failure to note particular matters should not be taken as approval of all portions of the statement not specifically discussed.

Essentially, the statement (other than the section on public broadcasting) makes two assumptions: (1) that there is widespread dissatisfaction with commercial television and (2) that broadcast facilities are scarce and then proceeds to the conclusion that cable and pay television are the answer. Both of the assumptions are fallacious, and the conclusion fails to take into account the essential public-interest questions in the cable-versus-free-broadcasting debate. I further am disappointed that the statement uses scarcity as the reason to defer recommending granting to broadcasting the full rights of the First Amendment until there are more voices through cable television; there are already many times more radio and television stations than daily newspapers.

Page 12, by JOHN A. SCHNEIDER, *with which* C. WREDE PETERSMEYER *has asked to be associated*

I dissent from the basic thrust and recommendations of this statement on national policy as summarized in this chapter and detailed in the pages that follow.

Broadcasting's contributions and successes are dismissed in a few sentences. Instead, a picture is presented of the American people rising almost as one, demanding more and better undefined efforts in the name of societal good. But one cannot honestly speak of "the public's expectations" as this document does. The voices of protest that are the loudest are those of the elitists who would substitute their judgments for those of the professionals and the public. So it is with this document.

In dealing with cable, this statement overlooks the trail of broken promises and virtually ignores the unfair competitive advantage that cable already has over free television and would give cable still greater leverage. Wittingly or not, these recommendations, if they could be completely implemented, would result in the replacement of free television, not with a better system, but with a pay system.

I also find it startling that this statement accepts government regulation of the broadcast press as a given. Thoughtful individuals are wondering about the wisdom of such regulatory involvement; the statement does not even ask the right questions.

There are, of course, commendable proposals in this statement. Unfortunately, some carry unworkable caveats. Equal time for political candidates would be lifted, for example, apparently to be replaced by wall-to-wall candidates (national, state, and local) appearing ad nauseum before elections.

Then there are the ringing proposals that lead to nowhere. The statement wants broadcasters "to establish identifiable goals and objectives that can provide a measure of success or failure in serving the public interest." That, of course, is what every broadcaster does every three years at license renewal time for all to see, and his community is the first to let him know whether he passed or failed. What more does the Committee want?

This policy statement is idealistic, as perhaps it should be. But idealism must converge with reality somewhere. This statement falls far short of that goal.

But the most unfortunate aspect of this document is that I came away not recognizing broadcasting as I and millions of viewers know it.

Page 12, by C. WREDE PETERSMEYER

I strongly disagree with the implication that such questions are widely held and reflect a general public discontent with broadcasting.

A medium as pervasive in its reach and impact as television, which tries to respond to the interests and tastes of 213 million Americans, is not going to satisfy every person all of the time. Moreover, like all of our institutions, both public and private, television is not perfect. It has had its failures, and it has its problems. But much of the criticism comes from those who fail to understand or are unwilling to accept the essential nature of the medium.

Commercial television is built on reaching most of the people most of the time with entertainment and information (hence the word *"broad*casting"). It is far and away the most broadly based of all the popular media. Providing a service that has such broad appeal constitutes a real and important service in the public interest. I disagree with those who seem to believe that the public interest is served only by limited-appeal programming of an educational or cultural nature.

Critics of television programming are more often than not those whose personal preferences do not coincide with what most people enjoy. Those critics want more of what they like, and they look with disdain on what most of their fellow citizens like. The fact is that the great majority of people prefer football to the ballet and Archie Bunker to Hedda Gabler. I do not find that fact upsetting.

Specialized program tastes have always been served to some extent by commercial television, and to a greater extent than is generally realized. Some regular programs and all sorts of specials attest to that. Whether such is enough is a matter that can be argued endlessly. Public television is, of course, designed to meet the demand for specialized programming on a full-schedule basis. The size of the demand is measured by the small share of audience (less than 3 percent) that regularly watches educational stations.

One indication of the regard in which the public holds television is found in the results of the most recent national opinion poll (fall 1974) conducted by the Roper Organization, which show that the public rated television stations higher, in terms of "excellent" and "good" performance, than local schools, government, newspapers, and churches. Moreover, television's performance rating is currently higher than at any time since such a question was first asked in 1959.

Page 13, by C. WREDE PETERSMEYER

Broadcasting facilities are not all that scarce. There are 947 television stations (93 percent of the public can receive four or more stations) and 7,715 radio stations. Broadcasting is a highly competitive business. If more channels were available, there is serious doubt that the economics of radio and television would ensure more stations than there now are, just as there are economic limitations on the number of newspapers, where spectrum limitation is not a factor. Good programming is expensive, and its cost must be amortized over a large number of viewers to support it.

The assertion that broadcasting occupies choice portions of the spectrum while new nonbroadcast users are crowding the spectrum faster than technology is finding ways to accommodate them fails to take into account the enormous inefficiencies in nonbroadcast use of the spectrum, the failure of the Federal Communications Commission to obtain an adequate data base on existing nonbroadcast uses, and its further failure to require effective utilization of frequencies now used by such users. Moreover, there is presently allocated for nonbroadcast use a large portion of the spectrum that is not being used at all.

Page 14, by C. WREDE PETERSMEYER

I see no connection between "enlarging the reach and impact of the media" and "encroaching on individual freedom and privacy."

Page 17, by C. WREDE PETERSMEYER

It is difficult to quarrel with the proposition that fair competition among technologies in the marketplace should be encouraged. But to urge that cable should be allowed to prove its value in the marketplace and to talk of encouraging fair competition between cable and television broadcasting is to pose, not to answer, the issues. Competition from "pure cable" (which would rely wholly on its own product, would carry no broadcast signals, would bid in the marketplace for its product, and would be subject to the same degree of copyright liability as broadcasting), of which there is virtually none, would be true marketplace competition. But competition with cable that uses broadcast signals to get into the home is neither fair nor true marketplace competition when such cable either pays no copyright fees for the regular television broadcast programs it purloins or receives a compulsory license on some fixed-fee basis or with arbitrators fixing the fees for the product, as to which the broadcaster could receive no such favorable treatment. The essential issue here is the extent to which cable should be permitted to import and hence sell the programs of free broadcasting, thereby building its business on the back of a business with which it expects to compete. (See my comments re copyright, page 96.)

Pages 17, 21, and 68, by ROBERT R. NATHAN, *with which* C. WREDE PETERSMEYER *has asked to be associated*

It is certainly desirable to rely to the maximum feasible degree on the marketplace, but it must be recognized that cable today relies primarily on rebroadcasting programs available from over-the-air broadcasts. Since cable must for the foreseeable future continue to rely heavily on over-the-air broadcasts, it would be most unfortunate if it were allowed to attract away from over-the-air broadcasting such programs as movies and sporting events that are now available without charge to viewers. This could weaken over-the-air broadcasting and also could compel people to pay for programs that they now view without direct charge.

There is no assurance that, if cable were to build viability by shifting current free over-the-air programs to a pay basis via cable, the cable stations would spend more money to develop new programs.

Every encouragement should be given to cable because it can become an important medium, but except for areas where over-the-air broadcasts are not readily received, cable is really a new industry that has not yet developed the economic foundation for its growth and development. Most important, that foundation should not be allowed to affect over-the-air broadcasting adversely. If that were to take place, both over-the-air and cable systems could be damaged.

Page 18, by OSCAR A. LUNDIN

The specific recommendations contained in this policy statement appear to me to represent reasonable steps toward achievement of an effective national telecommunications system. I recognize, however, that this is a highly technical subject and that the implementation of some of these recommendations may pose problems not readily evident to those who are unfamiliar with the complexities of the industry.

Pages 19 and 32, by C. WREDE PETERSMEYER, *with which* R. HEATH LARRY *and* HOWARD C. PETERSEN *have asked to be associated*

Safeguard from whom? The tentativeness of the statement's recommendations in not abolishing the fairness doctrine and not fully repealing the equal-time provision of the Communications Act until there are more voices (presumably cable) is based on a presumption that there is not now sufficient diversity of broadcast channels. There are already 8,662 broadcasting stations, compared with only 1,752 daily newspapers. In television alone, there are 136 cities with three or more television stations, compared with only five cities with three or more daily newspapers. Those who are dissatisfied with television programming for one reason or another have seized on the scarcity theory as the philosophical justification for the intrusion of government into the program process.

Why must broadcasting wait to have the full rights granted the "press" under the First Amendment almost 200 years ago! The First Amendment, prohibiting as it does government interference with communication among the people, is the cornerstone of all our freedoms, the rock on which our democratic society is built. That its protection should not have been *automatically* extended to television and radio is a grotesque paradox. What can be the meaning of the First Amendment if it is not to be applied to the most accepted form of communication we have. If Marconi had been born before Madison, is there any doubt as to what the Founding Fathers would have said?

I had hoped that the trustees of CED would have been on the battlefront along with other distinguished citizens who are urging the extension to the electronic press of the same freedoms granted the printed press. Former Senator

Sam J. Ervin, Jr., has characterized the enforced fairness concept as "a fickle affront to the First Amendment. . . . If First Amendment principles are held not to apply to the broadcast media, it may well be that the Constitution's guarantee of a free press is on its deathbed. . . . The broadcast media enjoy under the Constitution the same basic freedoms of expression that the newspaper does." Senator William Proxmire, who is waging an individual campaign on the Senate floor to extend First Amendment protection to broadcasters, has said, "When the Fairness Doctrine is examined for what it is, it is not fair at all. It is a form of prior restraint which does not square with First Amendment freedoms." And Justice William O. Douglas has said, "I fail to see how constitutionally we can treat TV and the radio differently than we treat newspapers. . . . The Fairness Doctrine has no place in our First Amendment regime. It puts the head of the camel inside the tent and enables administration after administration to toy with TV or radio in order to serve its sordid or its benevolent ends."

Pages 19 and 32, by C. WREDE PETERSMEYER, *with which* R. HEATH LARRY *has asked to be associated*

If the fairness doctrine must be maintained, I disagree with this recommendation that the FCC rule on fairness complaints promptly. Case-by-case review greatly intensifies the temptation by the FCC to second-guess the broadcaster. I would urge that the FCC judge the possible unfairness of a broadcaster in handling controversial issues only at license renewal time, based on a pattern of behavior over a longer period.

Pages 19 and 32, by C. WREDE PETERSMEYER, *with which* R. HEATH LARRY *has asked to be associated*

How can one "experiment" with fairness? What if some broadcasters *are* unfair? The Founding Fathers believed that the unchecked power of government was a far greater threat to liberty than the possible abuse of free speech by private citizens or organizations. I had hoped that the CED trustees would have agreed, without the necessity of "experiments."

Pages 19 and 34, by C. WREDE PETERSMEYER

Here again, the statement relies on the fallacious scarcity theory to delay the full repeal of Section 315 until there are more channels. CED was on the right track in its statement *Financing a Better Election System,* issued in 1968, when it recommended outright repeal of Section 315.

Pages 20 and 60, by C. WREDE PETERSMEYER

The statement talks in generalities about the twin goals of channel abundance and the resulting diversity of programming. Nowhere are the hard questions asked, let alone answered: Will an abundance of channels significantly increase the diversity of programming? If such diversity does come, who will benefit from it?

An abundance of channels, in terms of technological capability, does not guarantee program diversity. It is increasingly clear that cable, particularly pay cable, will seek to emphasize popular-appeal programming (like sports and movies), programming already the mainstay of free television. Moreover, if such diversity should become a reality, only those who are connected and who can pay the price will be able to take advantage of it. To the extent that the offerings of free television are diminished by subsidized cable, those who rely on free television (i.e., those who cannot afford cable and those who live in rural areas where stringing cable is uneconomic) are going to get less diversity and quality, not more.

Pages 21 and 68, by C. WREDE PETERSMEYER

Essentially, the statement proposes to subsidize the development of cable by continued use of broadcast signals and special copyright treatment because of the enormous capital cost necessary to develop a nationwide cable system. The statement regards this subsidization as important to achieve the "blue-sky" promises of broadband cable communications. Cable costs are indeed prodigious, as the statement says. But the statement presents no hard figures as to how prodigious those costs would be. Studies made for the President's Task Force on Communications Policy in 1968 indicate that, when adjusted to 1980 dollars, the capital costs of establishing a nationwide system of conventional cable would be approximately $231 billion and annual operating costs approximately $70 billion! Moreover, the costs of two-way video-grade service (referred to on page 13), which would be necessary to provide all of the "blue-sky" promises, would be astronomic, approximately $2 trillion in capital costs alone. It is difficult to see how this Committee can justify recommendations for profound changes in communications policy without more than the most superficial treatment of the actual costs that would be entailed.

If there is sufficient demand for cable's product (as the statement implies) that will "[widen] the range of programming and information services available," there is another way. Cable can make it on its own by creating special-interest programming, or indeed creating its own mass-appeal programming, *without* relying at all on retransmitting free broadcasting's product.

Pages 21 and 71, by ROBERT R. NATHAN

Common ownership of cable systems by broadcasters and networks should be allowed only under very strict rules and regulations. Perhaps only two cable channels should be available for use by such owners. It is very important that the broadcasters and networks do not control the very best programming and then argue that there is no demand for other channels. It will be necessary to establish regulations on rates charged for other channels to be sure that diversity of users and uses will not be hampered.

Pages 21 and 71, by C. WREDE PETERSMEYER

I oppose permitting broadcasters to own CATV systems in the same markets in which they own television stations. Allowing broadcasters to import signals that would compete against those of their own stations would inherently involve conflict of interest. There are hundreds of markets in which broadcasters can own and operate CATV systems, if they wish, other than those (a maximum of seven) in which they can own stations.

Pages 21 and 72, by E. B. FITZGERALD, *with which* CHARLES P. BOWEN, JR., *and* C. WREDE PETERSMEYER *have asked to be associated*

Unrestricted cable retransmission of distant television signals violates the "free market" right of program originators to undiminished enjoyment of their property. It is particularly troublesome to the professional sports industry that cable systems, which do not receive the permission of nor pay the sports program originators for the right to cablecast its sports events, are permitted to cablecast its games in unfair competition with the sports teams themselves and with broadcast television stations, which do pay for this right.

I concur with the policy recommendation to the extent that copyright protection must be increased in this area. However, I would add the requirement that any copyright proposals should specifically endorse the concept that a copyright proprietor in a free market should be able to control whether or not his programming material shall be carried at all as a distant signal on a cable television system.

Pages 21 and 72, by C. WREDE PETERSMEYER

Today, in the absence of copyright liability in any form for the retransmission of broadcast signals, cable stands in a unique and anomalous situation. It is akin to a retailer who takes whatever he wishes from wholesalers and manufacturers, sells the products for his own benefit, and never pays a dime to those who created or packaged those products. While the statement urges that major reforms in the copyright law are essential, it fails to come to grips with the basic elements of the copyright issue or to realize the implications of the copyright scheme it evidently endorses.

The type of copyright legislation for cable that is under active consideration today and that apparently is endorsed by the statement would provide some remuneration for those who create and own program rights but would do nothing to change the basic anomaly that exists today in the absence of copyright. The current copyright revision bills pending in Congress would adopt a scheme of compulsory licensing for cable retransmission of local broadcast signals and at least some distant broadcast signals. Compulsory licensing involves the payment of a specified fee, usually calculated in terms of a percentage of gross receipts. The policy statement simply asserts that if the parties are unable to agree on the amount of this fee, it could be established through compulsory arbitration.

Any compulsory license, regardless of the manner in which the amount of the fee is determined, would insulate cable from the competitive marketplace in the bargaining for program rights. The cable operator would still enjoy the enormous competitive advantage of simply taking whatever product was available to him under the compulsory license upon payment of the specified fee. Just as the automobile dealer, who could take all the automobiles he wished from each manufacturer upon payment of a nominal percentage of the gross receipts that the dealer obtained from retailing those vehicles, would enjoy an enormous competitive advantage over the dealer who must bargain both for a franchise and meet the manufacturer's price on each car, whether the dealer is ultimately able to sell the vehicle at a profit or not, so too a cable operator with a compulsory license enjoys an enormous competitive advantage over broadcasters who must bargain for the right to use programs.

If cable is to have such an enormous competitive advantage, it cannot complain on grounds of "free competition" when continuation of that competitive advantage is conditioned upon eschewing conduct which is inconsistent with maximizing both the quality and quantity of broadcast service. Specifically, cable cannot complain of limitations on distant-signal importation or on the types of programs that may be used for pay-cable purposes.

Pages 22 and 75, by CHARLES P. BOWEN, JR.

There are two fundamental issues here: (1) the feared impact upon an old, well-established information-distributing technology (over-the-air broadcasting) of competition from a newer one (cable) and (2) whether the older technology should be protected against cable by artificial, noneconomic operating restrictions imposed in broadcasting's behalf by a federal regulatory agency. Regulations limiting price competition for program material and for audience and limitations on program timing and format are not compatible with open-market competition and should be removed in the immediate future. Modernized copyright laws providing for negotiated fees to program owners are, of course, a parallel requirement.

Pages 22, 74, and 75, by EDWARD N. NEY

I believe a positive approach to cable regulation leads to these conclusions:

1. The reason I am concerned about commercial broadcasting is that it performs essential social functions relating to economic, political, and cultural life. It is a major community resource that must be maintained until its social functions can be adequately performed by other media.

2. I do not favor protectionist regulation that would keep new businesses based on new technologies from affecting the economic or social standing of existing businesses any more than I would have favored protecting horse-drawn carriages from the automotive industry or print from broadcasting or radio from television.

3. Broadcasters are in the business of packaging and distributing commercially sponsored programming. I believe that there will always be advertiser-supported television designed to reach large market segments and significant blocks of consumers. The programming may not be exactly what is broadcast today, but the economic and social needs will be there, and companies performing the function today will probably continue to do so. Broadcasting, after all, gets its name from its method of distribution, not from its product. The product and its manufacturers will remain in business even if distribution patterns change.

4. Under the best of circumstances, there will not be more than 1 million pay-TV subscribers by 1980. This is hardly enough to create the near-term threat to broadcasting that is often projected.

5. "Gradual relaxation of regulation" is impractical, as it never truly creates the option of free competition so that the results of such competition may be measured; nor does it offer any meaningful measure for the pace of relaxation, with the natural result that the pace will be dictated by political influence more than by justice.

6. Therefore, since I do not see the threat to broadcasting in the next five years projected by many, since I see nothing practical emerging from undefined gradualism, and since I favor free-market forces whenever possible, I recommend: (1) the total elimination of program-resource regulation from pay TV for five years and (2) an effort by regulators and social scientists during these five years to define the public-interest elements of broadcasting that should be protected as a national resource and measures for quantifying competitive threats to this resource so that at the end of five years we will have a better definition of our communications goals, experience, and measurable data upon which we will be able to define future policies.

Pages 22 and 76, by CHARLES P. BOWEN, JR.

With about 130,000 pay-cable subscribers out of 66 million TV homes, antisiphoning regulations are akin to protecting an elephant's feeding rights against interference from a mouse.

Pages 22 and 76, by E. B. FITZGERALD

Current "antisiphoning" restrictions are based upon the belief that unrestricted pay cable may lead to a shifting of sports events formerly shown on free, advertiser-supported television to pay cable. It must be pointed out, however, that this so-called siphoning threat is entirely speculative. Because of the novelty of pay-cable technology, there is simply not enough reliable market data to make sensible predictions and decisions on how the development of pay cable will affect broadcast viewing patterns or, for that matter, professional sports' gate receipts, the value of over-the-air broadcast packages, advertiser response, and finally, the financial health of broadcasters, professional team sports, and the cable industry. To fill this information void, teams should be permitted to offer the rights to otherwise untelevised games to pay-cable systems on an experimental basis. This will permit development of pertinent data upon which knowledgeable regulatory policies can be based.

The sports programming presented on pay cable should only be supplemental to that already available on free, over-the-air television. In this regard, it is interesting to note that the commissioner of baseball has, on numerous occasions, assured the FCC that there are no foreseeable circumstances under which such traditionally free television events as the World Series, the league championships, or the all-star games would ever be marketed on a pay-cable or subscription television basis.

Pages 22 and 76, by C. WREDE PETERSMEYER

While apparently conceding the undesirable social implications of permitting siphoning, at least with respect to sports events, the policy statement avoids the critical issues and alludes to several secondary, if not extraneous, issues. Pay cable is endorsed not merely as a means of bringing to the public special-interest programs which free television cannot provide but also, or even more importantly, as a means of stimulating cable growth. The possibility of the erosion of free television's audience and the financial base of broadcast television is mentioned as though these matters, rather than loss in free-program quality, were the primary issues. Since the purpose of the rules is to ensure that the public is not forced to pay for programs which it would otherwise have available free of charge, there is no point in "experimenting" with rules to determine whether there will be "unfair competition or other developments injurious to the public interest," whatever that may mean.

The statement is also quite unclear as to what sort of "experiment" is contemplated. Eliminating the rules in only a few selected communities would hardly provide a laboratory test of the incentives program suppliers would have to withhold their wares from free television if the rules were uniformly eliminated. On the other hand, a uniform, albeit "gradual and selective," weakening of the rules that led to undesirable consequences simply would not be remedied by a reimposition of the current rules *after* new investments in pay cable had been made. Time and again the FCC has found it impossible to impose controls retroactively. In its 1972 cable rules, for example, the FCC even "grandfathered" unbuilt cable systems that had a paper authorization to carry more broadcast signals than were permissible under the new rules on the theory that the FCC might otherwise disturb the expectations and investment *plans* of those holding paper authorizations. In the present context, those who rely on the *possibility* of reimposing rules as a justification for removing such controls for the present bear a heavy burden of showing how, as a practical matter, controls could be reimposed in the face of claims that multimillion-dollar investments in pay-cable hardware made in the interim would be jeopardized by reimposition.

If there are those who join in the policy statement and who believe that debasing the quality of free television service through pay-cable siphoning, at least in the case of motion pictures, is to be condoned in the hope that someday, somehow cable will ripen into something else, they ought in fairness at least to make it clear just what they are asking the American public to give up in the hope of getting something better.

Pages 22 and 81, by C. WREDE PETERSMEYER

I believe the statement is overly harsh on the FCC. The FCC's policies *have* resulted in the United States achieving the preeminent position in the world in communications (i.e., radio, television, satellite, and common carrier). Given the structure of government, the rapidly and constantly changing technology and the complexity of the issues involved in formulating policy that is in the public interest, the FCC has done a good job.

Page 28, by JOHN A. SCHNEIDER, *with which* CHARLES P. BOWEN, JR., HOWARD C. PETERSEN, *and* C. WREDE PETERSMEYER *have asked to be associated*

I dissent from the recommendations concerning government regulation of the broadcast press. This chapter recognizes the value of someday freeing the broadcast press from government regulation but postpones this deregulation until we move "from an era of technological scarcity to an era of abundance." Whatever is meant by those terms, I believe that the benefits to be obtained by the public from a free, vigorous, and independent broadcast press are too important to postpone to some future day.

I would have hoped that any examination of government regulation of the broadcast press—post Watergate—would have focused on the fundamental inappropriateness of having a governmental agency, however well intentioned, pass on the work of broadcast journalists. In affirming the need for a governmental presence to review the performance of the broadcast press, the statement appears merely to rely on the existing state of the law. The Committee nowhere explains why the present law is desirable, although many learned men in the courts and Congress are questioning governmental involvement in broadcast content.

If broadcast journalism is to be regulated because of the lack of station and network "abundance," is not such government regulation equally desirable with respect to the national wire services and major metropolitan newspapers? If not, why not? Such governmental regulation would be undesirable even though there are only two national wire services and only a handful of cities with competing newspapers. Only last year, the United States Supreme Court held a state law that imposed reply obligations on newspapers to be in violation of the First Amendment, despite contentions that economic factors made entry into the newspaper market impossible. It has, in my judgment, yet to be demonstrated why the policies supporting these First Amendment guarantees are not equally applicable to broadcast journalists.

For the same reasons that the *New York Times* and the *Washington Post* should not have to justify the "fairness" and "accuracy" of their investigative reporting before governmental agencies, I submit that CBS should not have to defend "The Selling of the Pentagon" before a congressional committee and NBC should not be forced to litigate the fairness of its broadcast

"Pensions: The Broken Promise." Yet, that is where we are today and where this statement proposes to leave us. To me, the answer is simple: If the broadcast press is to be truly free, its daily performance cannot be subject to government examination and approval—or disapproval.

This position is not mine alone. In a landmark decision less than two years ago, Supreme Court Justice Stewart wrote: "If we must choose whether editorial decisions are to be made in the free judgment of individual broadcasters, or imposed by bureaucratic fiat, the choice must be for freedom" (CBS v. Democratic National Committee, May 29, 1973). Obviously, some of my fellow trustees do not agree.

Given this statement's underlying philosophical premise, it is not unexpected that its conclusions involve the retention of the fairness doctrine, mandated free time for political candidates, and a general elitist assertion that television is not adequately meeting its social responsibilities.

Given my position at CBS, it will come as no surprise that I disagree with this pessimistic appraisal. I base my optimism on the fact that the ultimate arbiters—American viewers—have again and again found our broadcast product to be worthy of their trust and attention. If my fellow trustees know of a better test for the worthiness of a product, I welcome them to articulate it.

Page 37, by C. WREDE PETERSMEYER

I believe this substantially overstates the present situation. Because it is in almost everyone's home most of the time, television has never lacked for public response. Moreover, its high visibility invites attention from all sorts of individuals and groups, some well meaning and some not, some well informed and some not, some accepting the underlying premise of advertising and the competitive profit system and some not.

The problem for broadcasters in dealing with this is compounded by the great public diversity on matters of morality and taste, by radically changing standards over the last twenty years, and by the highly subjective and strongly held nature of individual judgments on these matters. The current vogue of consumerism and attendant group pressures add to the problem. I know of no industry that faces such a difficult task.

Sex and violence in programming, for example, are problems that broadcasters have wrestled with for years. The portrayal of conflict is the central theme in most drama. Sex has always been a major theme in most comedy. Keeping their treatment within reasonable limits in a world of changing values is a delicate and difficult task. Broadcasting may not have succeeded in every case. I would not try to defend every line that has been drawn. I can only say that broadcasters—as communicators, as citizens, and as parents—are sensitive to the problems and have tried hard to make careful decisions.

Page 38, by C. WREDE PETERSMEYER

This statement of belief (with which I wholeheartedly agree) that "the licensee must retain full control over [programming] decisions" does not square with other recommendations in the statement dealing with maintaining the fairness doctrine and the bulk of Section 315.

Page 41, by C. WREDE PETERSMEYER

These comments regarding the Television Code simply are not in accord with the facts. First, the code is enforceable through expulsion, and that is a sanction which most broadcasters today would regard with concern. Second, some subscribers have resigned when faced with the threat of expulsion. Third, the amount of commercial time is carefully monitored, and violations are brought to the attention of offending stations. If they do not fall in line, they are given the opportunity to resign or be expelled. Finally, in regard to violence and similar problems of program and commercial content, the code sets the standards. The networks in programming and the agencies in commercials apply those standards with considerable expenditure of money and effort. This is a very difficult area because of the wide diversity of viewer standards and the subjective nature of individual judgments. Honest disagreement with the decision on a particular program or commercial is certainly possible, but the mechanism is there, and it is used.

Page 41, by C. WREDE PETERSMEYER, *with which* CHARLES P. BOWEN, JR., *has asked to be associated*

The call for other "voices" to "contribute to the establishment of a meaningful and workable course for broadcasters," is in my judgment unwise. Broadcasters have always, by the nature of their business, listened to more people and more groups more carefully than any industry I know. It is besieged by voices. The institutionalization of the process is not needed. At best, it would be awkward and confusing; and at worst, destructive. Moreover, if such a procedure makes sense for broadcasting, why not have it for other industries like utilities, automobiles, oil, and newspapers?

Page 51, by C. WREDE PETERSMEYER

The notion that any significant number of additional VHF public television stations is feasible, other than by purchase, is totally unsupported. There is no information before the Committee to justify the suggestion that other VHF channels could be added to the spectrum without causing enormous loss of service to the public by interference with the signals of existing public and commercial VHF television broadcasting stations.

Page 60, by JOHN A. SCHNEIDER, *with which* C. WREDE PETERSMEYER *has asked to be associated*

I dissent from these recommendations for a national cable policy, recommendations that fail to recognize today's realities while fantasizing about tomorrow's world.

Cable is basically a parasitic industry, living off the investment and knowledge of others. To assist cable still further through the lifting of restrictions would only serve to penalize those who have made free television a viable industry. It would be costly to the American public and more costly to those who can least afford it. Lifting of restrictions without adequate protection would be the first step in the elimination of a free television system in favor of a paid system.

The statement fails to acknowledge that cable television now competes unfairly with broadcasting. There is competition between the two for audience, the lifeblood of both industries. The national television networks spend more than 50 percent of their gross advertising receipts for programming. Television stations spend about one-third of their receipts for local programming. Cable television's stock in trade is free broadcasting's programs, which cable retransmits and sells for a price. Cable television makes no contribution whatsoever to the cost of the retransmitted programs—with broadcasting paying the bill.

Cable, on the other hand, is completely at liberty to acquire exclusive rights against broadcasting to any event or program it chooses to originate, except to the extent that such origination is prohibited by the FCC. Broadcasting is not permitted to obtain exclusive rights to any event or program against general cable television retransmission. For example, a program broadcast on a television network in which exclusive rights are bought at high cost may be retransmitted by every cable television system in the United States— free.

I cannot agree that "cable's commercial success has been limited by an uncertain regulatory climate." The chart on page 65 speaks for itself. Cable television has grown extraordinarily in the economic and regulatory framework in which it has lived. Small wonder, because cable has enjoyed and continues to enjoy extraordinary advantages. But there is no reason that cable should have extraordinary guarantees of future growth. What is needed is the application of the normal law of copyright to cable, in which case there would be no need for many of the regulations that supposedly tend to limit the growth of cable. Unfortunately, this matter has not been kept in perspective by the statement.

The statement complains that present regulation of cable treats it as "an adjunct to broadcast television" and "an extension of broadcasting." This is realistic since cable is, and for the foreseeable future will continue to be, a dependent of broadcasting. Cable utilizes broadcast signals as its stock in trade and threatens, if importation of distant signals is not regulated, to destroy the broadcast industry upon which it exists. Part of the solution recommended by the statement is the application of copyright to cable.

The statement is marked by ambivalence. First, it urges full applicability of the copyright law, then notes that "determining just compensation for a creative product distributed by cable has been extremely difficult" and calls for economic studies to fix reasonable fees. Full applicability of the copyright law means that cable systems will have to enter the marketplace just as broadcasters do, and the marketplace will determine what the just compensation is. No economic studies should fix "reasonable" fee scales as a substitute for the free marketplace in copyright any more than such scales would be substitutes for the marketplace in any other field of endeavor.

The statement then advocates that the parties concerned should "jointly establish a schedule of royalty payments." No one who has looked into the matter can be unaware that such efforts in this field have broken down completely. As long as cable enjoys free use of the property of others, there will not be any agreement in this field.

The statement naïvely assumes that regulatory programming restrictions on cable television can be phased out and then reimposed if it "develops that pay cable is competing unfairly or harming the public interest." The history of regulation, including cable television, is a history of "grandfathering" because of the impossibility of turning back the clock.

Page 60, by C. WREDE PETERSMEYER

Although the statement here is careful to draw a distinction among "three distinct forms of cable," these distinctions are completely lost and obscured in the policy advocated later in this chapter. In essence, the statement urges that the "wide array of nonbroadcast entertainment and information services, including two-way communications," which are envisioned as the ultimate blue-sky future of cable, can and ought to be realized by stimulating the growth of a form of cable that is dependent upon the importation of distant broadcast signals and pay-cable originations of mass-appeal entertainment programs.

There is no reasonable basis for concluding that stimulating the growth of the kind of cable operation that is dependent on distant signals and pay-cable use of mass-appeal programs would lead to the development of the third "distinct form" of cable (i.e., true broadband communication). That this would come about is, of course, implied by cable entrepreneurs who are interested in marketing distant signals and pay cable in their "selling" effort in Washington and among opinion leaders. The very recent history of cable regulation shows the folly of assuming that cable operators would make the enormous additional investment needed to develop broadband communication. They have sought and have gotten more and more in the way of the relaxation of rules and have offered to do less and less in the way of ancillary services, once so important to their pitch. In 1972, the FCC gave cable growth a very significant boost by relaxing their prior limitations on distant-signal importation. Yet, there is no indication that the subsidization of cable growth through permissive distant-signal rules has brought the broadband communications "blue sky" any closer to realization today than it was in 1972. Indeed, the aspects of the FCC's 1972 regulatory program designed to compel, albeit in a very modest way, the development of new and innovative cable services are now being abandoned.

The requirement that larger CATV systems originate programs has been eliminated. The requirement of separate public, educational, and governmental access channels has been repeatedly waived by the FCC, and proposals to weaken or even eliminate those requirements are under consideration. In at least one case, the FCC has even found it difficult to force a CATV system to expand channel capacity sufficiently to deliver all of the local television broadcast stations that the system is obligated to carry.

Page 62, by C. WREDE PETERSMEYER

The policy was designed, *not* to protect over-the-air television service from unfair competition, but to *protect the viewer* from the threat of diminished free service.

Page 62, by C. WREDE PETERSMEYER

The existing access channels on many post-1972 cable systems have, with few exceptions, gone virtually unused. Cable operators vigorously opposed and were ultimately successful in eliminating the FCC requirement compelling program originations by larger cable operators, not because the cable operators had other or better uses for those channels, but because they argued persuasively that the incremental cost of filling an otherwise unused channel was uneconomic.

The obvious fact is that the big-market cable entrepreneurs (not the hundreds of small-town cable operators) want a part of the main action: mass-appeal programming. And they want to put it on a pay basis. Start with what is not now available on free TV (e.g., some home games and some first-run movies), and move out from there bit by imperceptible bit. They know that their economic future does not lie in special limited-appeal programming or exotic services except to the extent that talking about them enlists support from opinion leaders and politicians. They know that the move to a broadband communications system on anything approaching a nationwide basis is a quantum leap into an unknown future with costs of such magnitude, return on investment so speculative, and benefits to be gained so exotic compared to real world needs as to rule out development by private funding. Further, they know that government funding on so massive a scale would raise very basic questions of priorities in view of more pressing needs of society.

Page 63, by HERMAN L. WEISS, *with which* MARVIN BOWER *has asked to be associated*

I do not fully agree with this statement. While it is true that costs have risen beyond expectations, this is true in almost every business and is the direct result of double-digit inflation. The statement infers that, as a result, all potential subscribers have become disillusioned. I think it is a fact that very few cable television systems have had significant problems, outside of a few of the major metropolitan areas, in attracting and holding their subscribers. The decline in profitability and the scarcity of venture capital is a problem plaguing all businesses in general, and the basic reasons for this have little to do, in my opinion, with the technical progress of cable television. The fundamental reasons for the profitability trend and shortage of venture capital lie elsewhere.

Page 66, by C. WREDE PETERSMEYER

Included in the goals, indeed the first goal, should be protecting the viewer from diminution of his present free broadcasting service.

Page 68, by C. WREDE PETERSMEYER

The statement asserts that the "FCC's authority over cable is far from clear." There is no serious question as to the FCC's broad authority to regulate any cable system that retransmits broadcast signals, including any effect this kind of cable may have on the service television broadcast stations provide the American people. There may be some question as to the commission's authority with respect to cable systems that do not retransmit broadcast signals, of which there are virtually none.

Page 68, by CHARLES P. BOWEN, JR.

This says in effect that the only way freedom of program content can be achieved is by the power of wealth and that until cable reaches such a state the FCC should control its program content.

Page 69, by CHARLES P. BOWEN, JR.

We will never have enough information to persuade the proponents of strong regulation and protectors of the status quo that it is time to let the potential competitors risk their own assets trying to build pay-TV cable systems if they wish to do so. Pleas to postpone the decision appear to be based on the unstated assumption (which has yet to be justified in any complex techno-economic area) that regulators are likely to possess the knowledge, objectivity, and judgment that qualified them to substitute their decision making for that based on the risk takers' appraisal of the competitive technology and economics of the old and new systems.

Page 73, by CHARLES P. BOWEN, JR.

This is unfortunately typical of the kind of regulatory hairsplitting that occupies the agencies; avoids the basic issues; keeps the risk-reward situation uncertain; limits the rights of the film, sports, and other program owners; limits competition to the advantage of older technologies (i.e., movies and over-the-air broadcasting); and wastes the taxpayers' money.

Page 74, by C. WREDE PETERSMEYER

In contrast to the uncritical embracing of the belief that the more traditional forms of cable television would evolve into sophisticated broadband communications networks if rules with regard to distant signals and pay cable were weakened or eliminated, the statement expresses skepticism that free television service will be diminished in the absence of FCC rules on pay cable. While it may be impossible to predict the future with precision, past experience and common sense offer considerable insight into the likely development of pay cable in the absence of effective regulation. Pay cable is promoted, after all, as the box office in the home, and there has been ample opportunity over the years to observe the economics of the box office. (1) Entrepreneurs using the box-office method of distribution go to some length to control or curtail other access to the event. Walls are built around stadia and theaters that charge admissions. Home games are "blacked out" on local television. Motion pictures in theatrical exhibition are withheld from free television exhibition until the box-office distribution has run its course. Even radio and blow-by-blow television coverage of championship boxing are restricted by promoters using pay TV in theaters. These steps are taken to maximize revenues because many people will not pay if they can see the event free of charge. (2) Profit maximization in the box-office media dictates using mass-appeal, popular events whenever possible. The owner of a stadium would rather have a Super Bowl than a croquet tournament. (3) With only a relatively small number of homes or attendees, the pay or box-office medium will generate far more revenue with which to pay the entrepreneur than the advertiser-supported medium can generate.

Given these facts, it is highly probable that even with a relatively small number of pay-cable homes, the owners of popular motion pictures and sports events would have powerful incentives to withhold events from free television in order to exploit the new pay media. Unlike theaters and other pay media, however, the ability of pay cable to attract popular programming away from free television arises only because cable was able to exploit broadcast signals to get into the home in the first place. In this situation, there is nothing unfair or undesirable in conditioning cable's continued exploitation of broadcast signals upon compliance with rules designed to prevent the siphoning or shifting of programs from the free medium of near-universal availability to a pay medium of very restricted availability. That is all the FCC's rules purport to do.

Pay television, both over the air via subscription television stations and more recently through the cable medium, has traditionally been defended as a means of bringing to the public programming of relatively limited appeal, which advertiser-supported television is unable or unwilling to provide to the public. The present FCC rules permit the use of pay cable for these purposes. Live theater, ballet, opera, unusual sports, and all manner of musical productions as well as certain feature films and other sports events are available for pay-cable purposes under the present rules, and pay cable has come into existence and grown at a rather rapid rate in compliance with those rules. Thus, the issue is, *not* whether there should be a mix of "various forms of financing" to support programming via cable, but whether the public should be forced to pay or be deprived of programs that, except for the existence of pay cable, would otherwise be available on free television.

Page 80, by JOHN A. SCHNEIDER, *with which* C. WREDE PETERSMEYER *has asked to be associated*

I dissent from the recommendations for a reorganization of the regulatory system, which propose a course of action that we should all approach with caution.

The first warning signal comes from the need to reach back to the 1930s for support for the proposition that the Federal Communications Commission has not "come to grips with the major policy problems which are involved in the regulation of the radio industry." Since that time, of course, the United States has developed the most diversified, satisfying, and successful free broadcasting system in the world for both radio and television; the most efficient telephone system in the world; the technology for both domestic and international satellite systems; the ability to communicate from space probes deep into the solar system. All of this and more occurred in spite of what Chapter 5 refers to as the commission having "drifted, vacillated, and stalled in almost every major area."

It is clear what some of the critics of the commission cited by Chapter 5 were after: executive control of the administrative agencies. As applied to the FCC, that could mean executive control of the electronic news media.

Robert Cushman, the critic of the 1930s, proposed that administrative agencies be placed in executive departments, since he viewed the basic problem as their independence. James Landis, in a similar mood, proposed an office for the oversight of regulatory agencies to be created within the Executive Office of the President, to assure the efficient execution of the laws those agencies administer.

I would have thought we were too close to the recent attempts by the administration to muffle electronic news to see this suggestion raised anew. I also note that, contrary to the impression given by this chapter, the Hoover Commission found that the independent agencies had "largely achieved freedom from direct partisan influence in the administration of their statutes" and specifically rejected the Cushman conclusion that they be incorporated into the executive departments.

Both Cushman and Landis urged creation of a separate communications court to deal with adjudicatory matters in much the same fashion as recommended by Chapter 5. When first spoken, suggestions to "break up" the commission have a nice sound to them. As a life-long broadcast professional, I have felt the weight of commission indecision on significant matters. As a layman, however, I cannot agree that the adjudicatory process is any more precise or predictable. On the contrary, I find it a much more wasteful, dilatory, and expensive way of arriving at the same result—a decision that is firm and precise only until the next case is decided. In this regard, the Hoover report urged improving the hearing procedure; Hoover recognized the problem of costly adjudication and apparently saw fit not to suggest a separate communications court.

I had always thought Congress set the basic communications policy and the FCC was to administer it. We now have a suggestion that a further separation be created: Congress to set the basic policies, the commission to establish the rules and regulations, and a separate and independent communications court to interpret and apply those policies, rules, and regulations.

Unless this Committee is prepared to address directly whether the FCC should be abolished as an administrative agency (a course I am not suggesting), I cannot understand its urging creation of a separate communications court modeled after the Tax Court. For unless the commission continues to have adjudicatory functions, it would have no more justification for existence as an agency separate from the executive department than would the Internal Revenue Service, which issues tax regulations subject to adjudication before the Tax Court.

Page 83, by W. D. EBERLE

Although a court can review and adjudicate the issues before the commission, it does not create a method for greater responsibility by the FCC on public-policy issues. It is this responsibility to the public that is urgently needed and is really being addressed by the paper. In addition to the court, it might be well to give to the President and the oversight committees of the Senate and House the right to ask the FCC at any time to review a policy issue. Then, the FCC would hold public hearings and make a decision on these matters. This, I believe, would keep the regulatory agencies independent but, at the same time, responsive by making it appropriate and legal for Congress and the President to raise issues as needed. Obviously, Congress could do this by legislation at any time, but experience indicates prompter results can be achieved by Congress or the executive branch requesting action than by implementing positive action themselves, as they do not have the expertise or the time.

Page 86, by C. WREDE PETERSMEYER

This proposal ignores the realities of today. The present FCC is bombarded with views from all sorts of people and all sorts of organizations. In part this comes from commission encouragement over a number of years and in part from the increasing awareness of various groups that broadcasters and government agencies often respond to pressure. Abetting the process have been the public-interest law firms financed by various foundations.

As I see it, the danger today is quite the opposite. Vocal and well-financed special-interest groups usually get far more hearing and exert far more influence than their numbers justify. The great majority is seldom organized, almost never represented, and thus is seldom heard.

APPENDIX A

A BRIEF REGULATORY HISTORY OF CABLE TELEVISION

Cable television began in the late 1940s as community antenna television (CATV), a method of bringing existing television signals to mountainous or remote areas where reception was weak or nonexistent. A receiving antenna would be installed on a mountaintop; there, television signals were received, amplified, and sent by cable into homes in the valley below or other areas that had inadequate reception. Later, the idea spread to larger cities, where high buildings interfered with reception.

For years, the regulation of cable was left almost entirely in the hands of local government. Underfinanced, overburdened, and ill equipped to cope with new technologies, local governments saw cable mostly as a means of bringing improved television service to their citizens. They failed to appreciate the wider range of opportunities that the new technology might offer. Typically, a prospective cable operator, often the local television or hi-fi repairman, would come to a city council meeting requesting permission to run cables along and across city streets. An ordinance would be passed granting such a license. In some cases, cities would charge a franchise fee that went into its general fund; but in many cases, no other requirements were imposed on the operator. Many cities awarded franchises without even requiring that a system be built promptly; conversely, some systems operated without any franchise at all.

State regulations during cable's early growth might have brought some uniformity to the local franchising process, especially if the FCC had provided the states with policy guidance or at least with the assurance that state action would not be preempted by the federal government. Despite scattered attempts by state public utilities commissions or legislatures to assert control over cable, no state directly controlled cable television on the basis of a specific code until late in 1963, when Connecticut's General Assembly granted its Public Utilities Commission the power to award all cable franchises. Even today, only twelve states have assumed authority over cable, and only three have established separate cable commissions for this purpose.

As a result of the hands-off policy at the state and federal levels, hundreds upon hundreds of cable franchises were granted with requirements varying not only from state to state but from county to county and even from township to township. The crazy-quilt pattern that developed in local regulation of cable was as fragmented and as lacking in order as local governments themselves.[1]

The seeds of federal regulation of cable were planted in 1952 when the FCC lifted its freeze on new television licenses and disclosed a new plan for channel allocation. With the aim of furthering local service, the commission launched a major promotion of ultrahigh-frequency (UHF) television. However, the FCC's goal of gradually building a system of small-town stations conflicted with the public's desire to receive immediately at least the programming of the three networks. Many small communities could not support three network stations, and translators and boosters were widely used to relay signals beyond the normal coverage areas of big-city stations. Cable systems offered another alternative means of extending service.

At first, the FCC did not assert its jurisdiction over cable; and even today, the commission's authority over cable is not clear-cut. Congress has yet to enact any law that grants the FCC specific regulatory authority over cable. As a result, basic policies have been worked out by the commission under review by the courts.

Federal regulation developed indirectly; for example, the 1962 Carter Mountain Transmission Corporation case imposed certain carriage and nonduplication conditions on a microwave system transmitting signals to a cable system.[2] In 1965, these conditions were extended to all systems served by microwave, but cable systems not depending on microwave transmission were not subjected to regulation at that time.

During the sixties, cable grew in smaller communities. Its principal attraction, the importation of distant signals, began to disturb local broadcasters, who believed that cable was a form of unfair competition, that it would fragment their audiences and endanger their advertising revenues. Although cable was fulfilling a definite consumer need, the FCC also saw it as a threat to the local broadcasters and to the design for local service laid out in 1952. The FCC argued that if local stations were forced off the air, only cable subscribers would be able to receive television signals. Because of their cost, the FCC reasoned, cable systems would only serve densely populated areas whose residents were able to pay for service, thus leaving rural America and the urban poor with no television service at all.

---

[1] See *Modernizing Local Government* (1966).
[2] 44 FCC, 2776 (1962).

Therefore, in 1965, the FCC issued its First Report and Order, which extended its authority to the regulation of all microwave-fed cable systems.[3] Although the 1952 plan had ruled out government protection of broadcasters' markets, the First Report and Order argued that competition from cable must be limited in order to keep local service, especially UHF, alive. It contained two main requirements: Every cable system must carry the signal of every television station within approximately sixty miles, and it must refrain from carrying any program broadcast on any local station for fifteen days before and after that broadcast.

The commission's Second Report and Order was issued in March 1966, less than eleven months after the first.[4] It reaffirmed and asserted the FCC's jurisdiction over all cable systems and set down its new major market distant signal policy. This policy stated that cable systems in the top 100 markets could not import distant signals without express commission approval. Although cable systems could grow in small communities, the effect of this rule was that they could not bring additional programming into the top markets without a hearing. This restriction was based particularly on the FCC's concern over the future of UHF. The commission reasoned that UHF was most likely to grow in major metropolitan areas; thus, it dampened cable's growth in those areas.

Even after the Second Report and Order, there was some controversy about whether the FCC actually had the authority to regulate cable television. This was decided in part by the case of *Southwestern Cable Co.* vs. *United States,* which involved the importation of distant signals into a major market (San Diego). In 1968, the Supreme Court ruled that the Communications Act gave the FCC authority over cable television but that its authority was limited to what was reasonably ancillary to the effective performance of the commission's responsibility to regulate over-the-air television. The Court did not give an opinion on the FCC's power to regulate cable under other conditions or for other purposes.

As the commission acquired further experience with cable, it relaxed its policies. In 1968, it issued interim rules requiring larger cable systems to begin some new services (such as program origination) that had formerly been barred. It maintained the restriction on cable in the top 100 markets, although it reduced the protective zone around each central city substantially. The program-origination requirement was dropped in November 1974.

---

[3] 30 FCC, 683 (1965).
[4] 2 FCC, 2d 725 (1966).

These rules were in effect only until 1972, when the present rules governing cable were adopted. The Cable Television Report and Order (also known as the Third Report and Order) presented the most complex and comprehensive position taken by the FCC on cable.[5] Under these rules, cable was freer than before to expand into large markets, but the amount of competition it could offer was still limited. The FCC had modified its objective of protecting small stations and promoting UHF at cable's expense. However, it still restricted the expansion of cable into the larger markets, although it did not forbid such expansion.

Under the 1972 rules, all systems must carry all local stations, and distant signals may be imported only up to certain limits set by the FCC, depending on market size. The report stated that in comparison with systems in the top 100 markets, cable in smaller markets would be permitted to import fewer signals because the populations were smaller and because the small, local broadcaster would be more vulnerable to such competition. In order to open new outlets for local expression, the report stated that each system must provide channels for free public access, educational use, and local government use; it also specified standards of video and aural quality.

Before 1968, pay television (then known as subscription television, or STV) was authorized by the FCC only on an experimental basis. In 1968, the FCC issued regulations restricting the programming content of STV. In 1970, the FCC formally applied these rules to pay cable[6] and in 1972 incorporated them into the FCC's new rules on cable television. These regulations are: (1) Generally released feature films that are between two and ten years old may not be shown on cable television. Films that are over ten years old may be shown one week in each calendar month. (2) Sports events that have been on broadcast television in the past two years may not be offered by pay cable. (3) Serial programming is banned from pay cable. (4) Movies and sports cannot total more than 90 percent of the total pay cablecast hours, measured on a yearly basis, or 95 percent of the programming for any calendar month.

The FCC rules ban cable-system ownership within a particular community by broadcast television stations serving the same community; television networks are prohibited from owning cable systems. Telephone companies are also prohibited from operating systems within their telephone franchise area.

---

[5] 36 FCC, 2d 141 (1972).
[6] 23 FCC, 2d 825 (1970).

APPENDIX B CABLE REGULATIONS IN THE STATES

State	Systems [a]	Subscribers [a]	PUC/ PSC [b]	Cable Commission	Official Study Groups	Action Pending	Local Enabling Legislation
ALABAMA	75	175,791				X	
ALASKA	7	4,723	X				
ARIZONA	27	60,641					X
ARKANSAS	63	94,847				X	
CALIFORNIA	285	1,278,351					
COLORADO	37	70,951					
CONNECTICUT	3	25,781	X				
DELAWARE	8	57,281	X				
FLORIDA	108	416,572					
GEORGIA	69	207,918					
HAWAII	8	22,754	X				
IDAHO	42	60,218					
ILLINOIS	70	228,303				X[c]	
INDIANA	58	388,892					
IOWA	38	56,976					
KANSAS	77	135,748					
KENTUCKY	103	125,905					
LOUISIANA	33	78,139					
MAINE	29	43,902			X		
MARYLAND	27	83,479				X	X
MASSACHUSETTS	41	136,225		X			
MICHIGAN	64	212,187				X	
MINNESOTA	71	118,920		X			
MISSISSIPPI	57	133,513					
MISSOURI	59	118,643					
MONTANA	33	85,244					

STATE								
NEBRASKA	43	51,600						X
NEVADA	6	29,008	X					
NEW HAMPSHIRE	33	68,903						
NEW JERSEY	32	179,419	X[d]					
NEW MEXICO	28	78,427						
NEW YORK	156	634,114		X				
NORTH CAROLINA	36	122,946						
NORTH DAKOTA	10	19,190						
OHIO	137	439,152						
OKLAHOMA	75	130,179						
OREGON	91	163,345				X	X	
PENNSYLVANIA	300	952,781						
RHODE ISLAND	1	3,125	X					
SOUTH CAROLINA	33	59,782			X			
SOUTH DAKOTA	17	29,660						
TENNESSEE	57	109,055						
TEXAS	219	592,470				X		
UTAH	7	5,694						
VERMONT	36	49,713	X[d]					
VIRGINIA	54	122,419		X[e]				
WASHINGTON	101	236,334						
WEST VIRGINIA	134	231,737			X			
WISCONSIN	64	113,077				X		
WYOMING	25	53,602						

a Television Factbook, No. 44 (1974–75 ed.).
b PUC is the public utilities commission; PSC is the public service commission.
c Pending court decision; all other pending action is legislative.
d Regulated by an office or division within the PUC or PSC.
e The Virginia Public Telecommunications Council has assumed some control over cable television by establishing minimum technical standards to ensure effective use by state educational broadcasters.

Objectives of the Committee for Economic Development

For three decades, the Committee for Economic Development has had a respected influence on business and public policy. Composed of two hundred leading business executives and educators, CED is devoted to these two objectives:

To develop, through objective research and informed discussion, findings and recommendations for private and public policy which will contribute to preserving and strengthening our free society, achieving steady economic growth at high employment and reasonably stable prices, increasing productivity and living standards, providing greater and more equal opportunity for every citizen, and improving the quality of life for all.

To bring about increasing understanding by present and future leaders in business, government, and education and among concerned citizens of the importance of these objectives and the ways in which they can be achieved.

CED's work is supported strictly by private voluntary contributions from business and industry, foundations, and individuals. It is independent, nonprofit, nonpartisan, and nonpolitical.

The two hundred trustees, who generally are presidents or board chairmen of corporations and presidents of universities, are chosen for their individual capacities rather than as representatives of any particular interests. By working with scholars, they unite business judgment and experience with scholarship in analyzing the issues and developing recommendations to resolve the economic problems that constantly arise in a dynamic and democratic society.

Through this business-academic partnership, CED endeavors to develop policy statements and other research materials that commend themselves as guides to public and business policy; for use as texts in college economics and political science courses and in management training courses; for consideration and discussion by newspaper and magazine editors, columnists, and commentators; and for distribution abroad to promote better understanding of the American economic system.

CED believes that by enabling businessmen to demonstrate constructively their concern for the general welfare, it is helping business to earn and maintain the national and community respect essential to the successful functioning of the free enterprise capitalist system.

Statements issued in association with CED counterpart organizations in foreign countries.

Further Weapons Against Inflation *(November 1970)*

Making Congress More Effective *(September 1970)*

*Development Assistance to Southeast Asia *(July 1970)*

Training and Jobs for the Urban Poor *(July 1970)*

Improving the Public Welfare System *(April 1970)*

Reshaping Government in Metropolitan Areas *(February 1970)*

Economic Growth in the United States *(October 1969)*

Assisting Development in Low-Income Countries *(September 1969)*

*Nontariff Distortions of Trade *(September 1969)*

Fiscal and Monetary Policies for Steady Economic Growth *(January 1969)*

Financing a Better Election System *(December 1968)*

Innovation in Education: New Directions for the American School *(July 1968)*

Modernizing State Government *(July 1967)*

*Trade Policy Toward Low-Income Countries *(June 1967)*

How Low Income Countries Can Advance Their Own Growth *(September 1966)*

Modernizing Local Government *(July 1966)*

A Better Balance in Federal Taxes on Business *(April 1966)*

Budgeting for National Objectives *(January 1966)*

Presidential Succession and Inability *(January 1965)*

Educating Tomorrow's Managers *(October 1964)*

Improving Executive Management in the Federal Government *(July 1964)*

Trade Negotiations for a Better Free World Economy *(May 1964)*

Union Powers and Union Functions: Toward a Better Balance *(March 1964)*

Japan in the Free World Economy *(April 1963)*

Economic Literacy for Americans *(March 1962)*

Cooperation for Progress in Latin America *(April 1961)*

Statements issued in association with CED counterpart organizations in foreign countries.